Sixth-Grade Math Minutes

One Hundred Minutes to Better Basic Skills

Written by
Doug Stoffel

Editor: Sue Jackson
Senior Editor: Maria Elvira Gallardo, MA
Cover Illustrator: Rick Grayson
Production: Rebekah O. Lewis
Cover Designer: Barbara Peterson
Art Director: Moonhee Pak
Managing Editor: Betsy Morris, PhD

TABLE OF CONTENTS

INTRODUCTION

The focus of *Sixth-Grade Math Minutes* is math fluency—teaching students to solve problems effortlessly and rapidly. The problems in this book provide students with practice in every key area of sixth-grade math instruction, including

- computation
- number sense
- reading graphs
- problem solving
- patterns and sequences
- data analysis and probability
- spatial reasoning
- fractions
- algebra and functions
- geometry

Use this comprehensive resource to improve your students' overall math fluency, which will promote greater self-confidence in their math skills as well as provide the everyday practice necessary to succeed in testing situations.

Sixth-Grade Math Minutes features 100 "Minutes." Each Minute consists of 10 classroom-tested problems of varying degrees of difficulty for students to complete within a one- to two-minute period. This unique format offers students an ongoing opportunity to improve their own fluency in a manageable, nonthreatening format. The quick, timed format, combined with instant feedback, makes this a challenging and motivational assignment students will look forward to using each day. Students become active learners as they discover mathematical relationships and apply acquired understanding to complex situations and to the solution of realistic problems in each Minute.

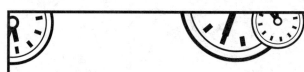

How to Use This Book

Sixth-Grade Math Minutes is designed to be implemented in numerical order, starting with Minute One. Students who need the most support will find the order in which skills are introduced most helpful in building and retaining confidence and success. For example, the first few times that students are asked to recognize rows and columns in a table, the particular row or column is shaded. Later the students are asked to recognize a particular row or column without the aid of shading.

Sixth-Grade Math Minutes can be used in a variety of ways. Use one Minute a day as a warm-up activity, bell work, review, assessment, or homework assignment. Other uses include incentive projects and extra credit. Keep in mind that students will get the most benefit from their daily Minute if they receive immediate feedback. If you assign the Minute as homework, correct it in class as soon as students are settled at the beginning of the day.

If you use the Minute as a timed activity, place the paper facedown on the students' desks or display it as a transparency. Use a clock or kitchen timer to measure one minute—or more if needed. As the Minutes become more advanced, use your discretion on extending the time frame to several minutes if needed. Encourage students to concentrate on completing each problem successfully and not to dwell on problems they cannot complete. At the end of the allotted time, have the students stop working. Then read the answers from the answer key (pages 108–112) or display them on a transparency. Have students correct their own work and record their scores on the Minute Journal reproducible (page 6). Then have the class go over each problem together to discuss the solution(s). Spend more time on problems that were clearly challenging for most of the class. Tell students that concepts that seemed difficult for them will appear again on future Minutes and that they will have another opportunity for success.

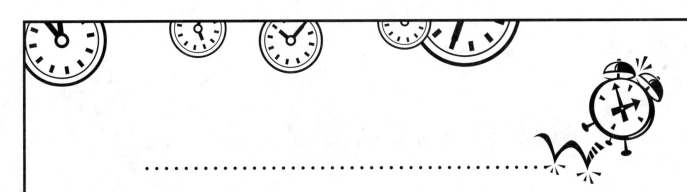

Teach students strategies for improving their scores, especially if you time their work on each Minute. Include strategies such as the following:

- leave more time-consuming problems for last
- come back to problems they are unsure of after they have completed all other problems
- make educated guesses when they encounter problems with which they are unfamiliar
- rewrite word problems as number problems
- use mental math whenever possible
- underline important information
- draw pictures

Students will ultimately learn to apply these strategies to other timed-test situations.

The Minutes are designed to improve math fluency and should not be included as part of a student's overall math grade. However, the Minutes provide an excellent opportunity for you to see which skills the class as a whole needs to practice or review. This information will help you plan the content of future math lessons. A class that consistently has difficulty reading graphs, for example, may make excellent use of your lesson in that area, especially if the students know they will have another opportunity to achieve success in reading graphs on a future Minute. Have students file their Math Journal and Minutes for the week in a location accessible to you both. You will find that math skills that require review will be revealed during class discussions of each Minute. You may find it useful to review the week's Minutes again at the end of the week with the class before sending them home with students.

While you will not include student Minute scores in your formal grading, you may wish to recognize improvements by awarding additional privileges or offering a reward if the entire class scores above a certain level for a week or more. Showing students that you recognize their efforts provides additional motivation to succeed.

MINUTE JOURNAL

....

NAME _____

Minute	Date	Score	Minute	Date	Score	Minute	Date	Score	Minute	Date	Score
1			26			51			76		
2			27			52			77		
3			28			53			78		
4			29			54			79		
5			30			55			80		
6			31			56			81		
7			32			57			82		
8			33			58			83		
9			34			59			84		
10			35			60			85		
11			36			61			86		
12			37			62			87		
13			38			63			88		
14			39			64			89		
15			40			65			90		
16			41			66			91		
17			42			67			92		
18			43			68			93		
19			44			69			94		
20			45			70			95		
21			46			71			96		
22			47			72			97		
23			48			73			98		
24			49			74			99		
25			50			75			100		

Sixth-Grade Math Minutes © 2007 Creative Teaching Press

Scope and Sequence

MINUTE 1

1. Circle the number that has a 4 in the tens place.   324   24   4,321   49

2. Circle the set of lines that are parallel.

3. Write these decimals in order from least to greatest.   0.403   0.034   0.340

_____   _____   _____

4. Write the fraction that represents the shaded boxes.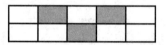

_____

5. $5 + \boxed{} = 12$

6. Complete the pattern: 1, 5, 9, 13, ____.

7. What is the area (number of squares) in the rectangle to the right?

_____

8. According to the chart, how many desks are in column A?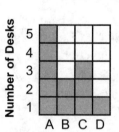

_____

9. $9 \times 4 =$
$9 \times 7 =$
$9 \times 9 =$

10. $7\overline{)28} =$ $7\overline{)42} =$ $7\overline{)63} =$

Sixth-Grade Math Minutes © 2007 Creative Teaching Press

MINUTE 2

1. If you flip a coin 10 times, how many times will it land on heads?

    a. 10   **b.** 5   **c.** 2   **d.** impossible to tell

2. Which shape is a pentagon?

    a.   **b.**   **c.**   **d.**

3. Write the fraction for each:

Two-fifths = _____

Three-fourths = _____

4. Write the fraction that represents the shaded boxes. _____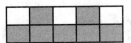

5. $3 \times 4 + 4 =$

6. Complete the pattern: 4, 8, 12, 16, _____.

7. What is the perimeter (distance around) of the rectangle to the right? _____.

8. According to the graph to the right:

    A = _____

    B = _____

    C = _____

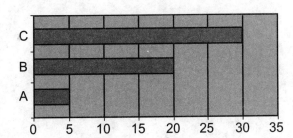

9. $8 \cdot 6 =$ $8 \cdot 4 =$ $8 \cdot 7 =$

10. $\dfrac{24}{6} =$ $\dfrac{36}{6} =$ $\dfrac{18}{6} =$

MINUTE 3

1. If it is 5:32 now, what time will it be 24 minutes from now? _____

2. How many cubes are in this shape? _____

3. Write two fractions that represent the shaded boxes.

_____

4. Write > or < in the circle to compare the fractions. $\frac{7}{9}$ ◯ $\frac{8}{9}$

5. Mel makes arm bracelets. She is making one for each arm of her six friends. How many should she make? _____

6. Complete the pattern. 2, 4, 8, _____.

7. Joe wants to build a fence for his dog Charlie. He plans to surround the rectangle to the right with fence. How many feet will he need? _____

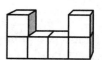

10 ft.

15 ft.

8. How many people took part in this survey?

_____

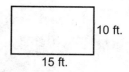

Favorite Cereals

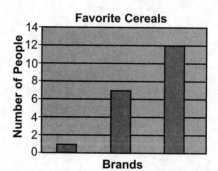

9. (12)(3) =
(12)(5) =
(12)(6) =

10. 50 ÷ 5 =   55 ÷ 5 =   45 ÷ 5 =

Sixth-Grade Math Minutes © 2007 Creative Teaching Press

MINUTE 4

1. Circle the number with a 5 in the tenths place.   36.05   41.5   50.313   15.38

2. Which of these shapes is a trapezoid?

   a. ▭   b. ⬠   c. ▽   d. ⬡

For Problems 3–4, write > , <, or =. Use the bars to help you.

3. $\dfrac{3}{6}$ ◯ $\dfrac{1}{3}$

4. $\dfrac{1}{4}$ ◯ $\dfrac{1}{3}$

5. $2(4 + 7) =$

6. Complete the pattern. 123, 234, 345, _____.

Justin's Garden

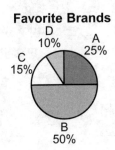

5 ft.

9 ft.

7. Justin has 30 feet of fence. Would this be enough to surround his garden? Circle: Yes   or   No

Favorite Brands

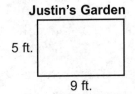

8. According to the chart, Brand B was chosen twice as often as Brand _____.

9. $1 + 2 + 3 =$
$3 + 4 + 5 =$
$5 + 6 + 7 =$

10.
   38   43   26
 + 37   + 96   + 57

MINUTE 5

1. The height of a room would most likely be 10 _____.
    a. feet   **b.** inches   **c.** yards

2. Which letter on the shape is beside a right angle? _____

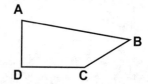

3. $\frac{1}{2}$ of 20 =

4. Write as a decimal: two and three-tenths = _____.

5. If the pattern continues, how many boxes should be shaded in row D? _____

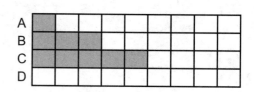

6. $(2 \times 3) + (3 \times 4) =$

7. What is the area of the shape to the right? _____

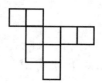

8. In the chart to the right, the y numbers are _____ times the x numbers.

x	1	2	4
y	3	6	12

9.   49   51
       − 28   − 32

10.   14   23
       × 5   × 7

Sixth-Grade Math Minutes © 2007 Creative Teaching Press

MINUTE 6

1. To build a school, it might take two _____.
    a. days   **b.** weeks   **c.** years

2. Which letter on the shape is beside an obtuse angle? _____

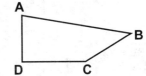

3. Which of the following is (are) equal to $\frac{1}{2}$?

    a. $\frac{5}{10}$ **b.** $\frac{7}{14}$ **c.** $\frac{10}{25}$ **d.** $\frac{12}{30}$

4. Write as a decimal: twenty-three hundredths = _____.

5. The library, post office, and gas station are all on Elm Street. The library is three miles west of the post office. The gas station is six miles east of the post office. How far apart are the library and gas station? _____

6. Complete the pattern. A12, B16, C20, _____, _____.

7. What is the area of a rectangle with a length of 9 feet and a width of 7 feet? _____

For Problems 8–9, use the bar graph to the right.

8. On what day of the week did Ron bowl the best? _____

9. On which two days of the week did Ron have the same score?

    _____ _____

Ron's Bowling Scores

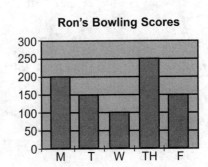

10.
11 + 43 =
26 + 19 =
18 + 17 =

MINUTE 7

1. Which of these shapes does not belong?

a.   b.   c.   d.

2. Which letter on the shape is beside an acute angle? _____

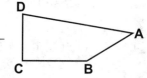

3. Which of the following is (are) equal to $\frac{1}{4}$?

a. $\frac{5}{20}$ b. $\frac{7}{21}$ c. $\frac{10}{40}$ d. $\frac{12}{50}$

4. Write as a decimal: Forty-three thousandths = _____

5. If $a = 10$ and $b = 6$, then $a + b = 16$.   Circle: True   or   False.

6. Draw the next shape in the sequence.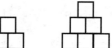

7. What is the perimeter of the shape to the right? _____

For Problems 8–9, use the chart to the right.

8. Which student had the best grade? _____

9. Desiree's score was about twice as high as the score for _____ .

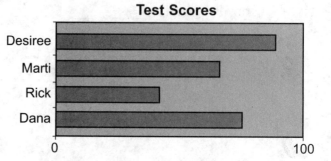

Test Scores

10. $3\overline{)636}$ = $3\overline{)129}$ = $3\overline{)501}$ =

Sixth-Grade Math Minutes © 2007 Creative Teaching Press

MINUTE 8

1. Justine's bill at a restaurant is $14.58. She pays with a twenty dollar bill. How much change should she get back? _____

For Problems 2–3, use the diagram to the right.

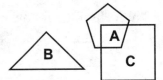

2. Which letter is inside the square and pentagon? _____

3. Which letter is outside the pentagon but inside the triangle? _____

4. Write the fraction for the shaded part in each figure below.

A. 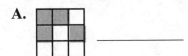 _____   B. 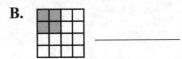 _____

5. If 7 out of 11 balloons are red, what fraction of balloons are NOT red? _____

6. Complete the pattern. 1, 2, 4, 7, 11, _____.

For Problems 7–8, use the bar graph to the right.

7. During which month(s) did more than 200 customers visit the store?

_____

Customers Per Month

8. In August, half as many customers visited the store as in _____.

9.
$$3.6 \qquad 4.9 \qquad 12.75$$
$$\underline{-\,0.7} \qquad \underline{-\,0.6} \qquad \underline{-\,0.35}$$

10.
$$22 \qquad 34 \qquad 46$$
$$\underline{\times\,4} \qquad \underline{\times\,5} \qquad \underline{\times\,6}$$

Sixth-Grade Math Minutes © 2007 Creative Teaching Press

MINUTE 9

1. Round each number to the nearest ten.

24 = _____ 311 = _____ 107 = _____

2. Which of the following shapes has a right angle?

a. ⬭ b. △ c. ◺ d. ⬡

3. Which of the following groups of numbers is in order from least to greatest?

a. 323, 411, 421, 506 **b.** 108, 106, 217, 304

c. 98, 94, 36, 29 **d.** 200, 199, 198, 405

4. Which of the following is NOT equal to 45?

a. $3 \times 10 \times 2$ **b.** $3 \times 3 \times 5$

c. $10 + 10 + 10 + 10 + 5$ **d.** $50 - 5$

5. $12 \times \boxed{} = 48$

6. Complete the pattern. $\dfrac{1}{2}, \dfrac{2}{3}, \dfrac{3}{4},$ _____.

7. Which shape has a greater area? _____

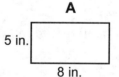

A 5 in. | 8 in.   B 3 in. | 12 in.

For Problems 8–9, use the chart to the right.

8. Which car weighs the most? _____

9. How much more does the red car weigh than the green car? _____

Weights of cars	
Color	**Weight in pounds**
Blue	2,786
Red	3,196
Green	2,500

10.
$$\begin{array}{r} 1.2 \\ \times\, 0.6 \\ \hline \end{array} \qquad \begin{array}{r} 1.4 \\ \times\, 0.7 \\ \hline \end{array} \qquad \begin{array}{r} 2.6 \\ \times\, 0.8 \\ \hline \end{array}$$

Sixth-Grade Math Minutes © 2007 Creative Teaching Press

MINUTE 10

1. Which of the following numbers is NOT equal to 36?

a. 4×9 **b.** $18 + 18$ **c.** $40 - 6$ **d.** $10 + 10 + 10 + 6$

2. Which one of these shapes has four vertices (corners)?

a.   **b.**   **c.**   **d.**

3. Which of the following groups of numbers is in order from greatest to least?

a. 323, 411, 421, 506   **b.** 108, 106, 217, 304

c. 98, 94, 36, 29   **d.** 200, 199, 198, 405

4. Complete the chart.

Add 0.4	
Start	**End**
2.2	2.6
3.1	
4.7	

5. $28 \div \boxed{} = 7$

6. Complete the pattern: $\dfrac{1}{3}, \dfrac{2}{5}, \dfrac{3}{7},$ _____.

7. Which shape has the greater perimeter?

_____

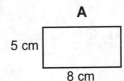
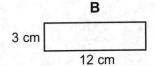

A   B

5 cm   8 cm   3 cm   12 cm

For Problems 8–9, use the bar graph to the right.

8. How many eggs did Lucky lay last season?

_____

9. How many more eggs did Clucky lay than Lucky? _____

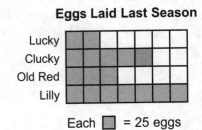

Eggs Laid Last Season

Lucky
Clucky
Old Red
Lilly

Each ▨ = 25 eggs

10.

$\begin{array}{r} 3.3 \\ + 2.4 \\ \hline \end{array}$ $\begin{array}{r} 4.5 \\ + 5.6 \\ \hline \end{array}$ $\begin{array}{r} 7.2 \\ + 10.3 \\ \hline \end{array}$

Sixth-Grade Math Minutes © 2007 Creative Teaching Press

MINUTE 11

1. Circle the number with a 4 in the thousands place.   324   421   4,321   49

2. Which of these shapes is a hexagon?

   a. ▭   **b.** ⬠   **c.** ⬡   **d.** ⬡

3. Which of the following is NOT equal to 40?

   a. $4 \times 8 + 8$ **b.** $2 \times 2 \times 5$ **c.** $10 + (5)(6)$

4. Put the fractions in order from least to greatest $\dfrac{3}{8}, \dfrac{7}{8}, \dfrac{2}{8}, \dfrac{8}{8}$. _____.

5. If $\dfrac{42}{x} = 7$, then $x =$ _____.

6. Complete the pattern: 12, 15, 17, 20, 22, 25, _____.

7. How many cubes would three layers of this shape have? _____

8. According to the graph to the right:

   A = _____
   B = _____
   C = _____

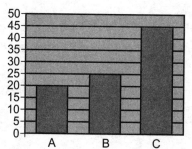

9. $9 \cdot 7 =$ $8 \cdot 8 =$ $6 \cdot 7 =$

10. $3 + 5 + 7 =$ $4 + 7 + 6 =$ $2 + 9 + 8 =$

Sixth-Grade Math Minutes © 2007 Creative Teaching Press

MINUTE 12

1. About how many commercials might have been shown this year during the Super Bowl?

    a. 4   **b.** 40   **c.** 400

2. Which letter on the shape is beside an obtuse angle? _____

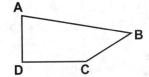

3. Which of the following groups of numbers is in order from least to greatest?

    a. 0.312, 0.411, 0.601, 0.806   **b.** 10.8, 10.6, 31.7, 40.4

    c. 0.88, 0.84, 0.76, 0.49   **d.** 5.00, 3.19, 1.98, 0.755

4. If $\frac{1}{4} = \frac{x}{8}$, then $x =$ _____.

5. Anna finished a race five yards ahead of Jack. Jack finished nine yards ahead of Tina. How many yards ahead of Tina was Anna? _____

6. Forty tickets were sold for a lottery. If Lon bought two tickets, what are the chances he will win? _____

7. What is the perimeter of the triangle? _____

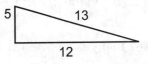

8. How many glasses of lemonade did Rhonda sell? _____

Glasses of Lemonade Sold

Justin	☺	☺	☺	☺	
Leah	☺	☺			
Rhonda	☺	☺	☺		
Candice	☺				

Each ☺ = 10 glasses.

9.   2.6   3.8

      + 3.2   + 4.5

10.   5.6   6.3

      × 10   × 10

NAME:

MINUTE 13

1. Round each number to the nearest hundred.

124 =   2,311 =   48 =

For Problems 2–3, use the diagram to the right.

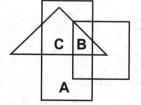

2. What letter is inside the triangle and the rectangle that is not in the square? _____

3. Which letter is inside of all three shapes? _____

4. Circle the fraction that is NOT in its simplest form.

$\frac{1}{4}$ $\frac{2}{5}$ $\frac{3}{8}$ $\frac{2}{6}$

For Problems 5–6, use the chart to the right.

4th Grade Classes		
	Boys	Girls
Room 1	12	13
Room 2	15	11

5. According to the chart, what fraction of the total number of students in Room 1 are boys? _____

6. How many boys are in Rooms 1 and 2? _____

7. $3 \cdot 4 + 2 \cdot 2 = 16$ Circle: True   or   False

8. A car salesman says he will give out a prize one day of next week to anyone who test drives a car. What is the probability that he will give out this prize on Thursday? _____

9. $\frac{1}{2} \times \frac{1}{3} =$ $\frac{1}{3} \times \frac{1}{4} =$ $\frac{1}{5} \times \frac{1}{6} =$

10.   46   79   88
     −16   −16   −16

Sixth-Grade Math Minutes © 2007 Creative Teaching Press

MINUTE 14

1. In the number 1,846, the _____ is in the tens place and the _____ is in the hundreds place.

2. Which of these shapes best represents a cube?

   a.   **b.** 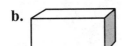   **c.**   **d.**

3. Circle the fraction that is NOT in its simplest form.

$$\frac{5}{11} \qquad \frac{5}{15} \qquad \frac{5}{12} \qquad \frac{5}{18}$$

4. If $\frac{2}{3} = \frac{a}{15}$, then $a =$ _____.

5. $\boxed{} + 11 = 20$

6. These four cubes were placed in a bag. What is the probability that the dark one would be pulled out of the bag first? _____

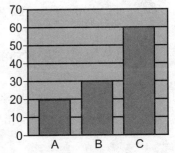

For Problems 7–8, use the bar graph to the right.

7. Which of the following statements is (are) true about the graph?

   a. A + B = 50   **b.** C is half of B   **c.** B is more than A

8. A + B + C is closest to: **a.** 50   **b.** 100   **c.** 200

9. Change to decimal form.

$$2\frac{1}{2} = \qquad 3\frac{1}{4} = \qquad 20\frac{1}{2} =$$

10. $\dfrac{20}{4} = \qquad \dfrac{30}{5} = \qquad \dfrac{40}{8} =$

MINUTE 15

1. What is the value in cents of 2 quarters, 3 dimes, and 4 nickels? _____

2. Circle the set of lines that are perpendicular: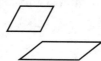

3. Which set of shapes shows two figures that are <u>congruent</u>? _____
   a.   b.   c.

For Problems 4–5, write > , <, or =.

4. $\dfrac{2}{8}$ ◯ $\dfrac{2}{9}$

5. $\dfrac{1}{5}$ ◯ $\dfrac{2}{10}$

6. Complete the pattern: 5, 7, 4, 6, 3, 5, ____.

7. What is the perimeter of a square if each side is 5 feet? _____

8. The y numbers in this chart are _____ times the x numbers.

x	y
2	10
3	15
7	35

9.
$$150 \quad\quad 275 \quad\quad 325$$
$$\underline{-\,25} \quad\quad \underline{-\,125} \quad\quad \underline{-\,75}$$

10. $5\overline{)155}$ = $4\overline{)408}$ =

Sixth-Grade Math Minutes © 2007 Creative Teaching Press

MINUTE 16

1. I have a 1 in the ones place, a 4 in the tens place, and a 5 in the hundreds place. What number am I? _____

2. Which letter is beside an acute angle? _____

3. Which set of figures shows two shapes that are similar but not congruent (same size and shape)?

  a.   b.   c.

4. Which fraction is in the simplest form?

  a. $\dfrac{5}{10}$ b. $\dfrac{7}{14}$ c. $\dfrac{10}{25}$ d. $\dfrac{12}{25}$

5. $3 + 5 + \boxed{} = 12$

6. Complete the pattern. 3, 5, 9, 11, 15, 17, _____.

7. What is the area of a rectangle that is 15 feet long and 3 feet wide? _____

For Problems 8–9, use the bar graph to the right.

8. According to the chart, which class has the same amount of boys and girls in it?_____

9. About how many more girls than boys does Class 1 have? _____

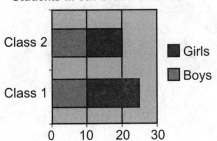

10.
$$3.8 - 2.6$$
$$14.06 - 1.01$$
$$10.0 - 6.5$$

MINUTE 17

1. Eileen's bill for her lunch was $7.33. She gave the waiter $10 and told him to keep the change as a tip. How much of a tip did the waiter get? _____

2. Which of these shapes best represents a cylinder? _____

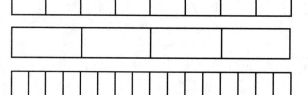

a.　　　　　　b.　　　　　　　c.

For Problems 3–4, write > , <, or =. Use the bars to help you.

3. $\frac{3}{8}$ ◯ $\frac{1}{4}$

4. $\frac{3}{4}$ ◯ $\frac{9}{16}$

5. $3 \cdot 2 + 6 \div 2 =$

6. Which shape has a greater perimeter? _____

A
4
8

B
2
11

7. A ball is dropped on the tiles to the right. What are the chances that it would land on a shaded tile? _____

For Problems 8–9, use the chart to the right.

8. Which student gets the largest allowance each week? _____

9. Which student gets $15 each week? _____

Allowances per Week				
Sandy	$			
Jared	$	$	$	$
Jackie	$	$	$	

$ sign = $5

10.

300	250	450
− 50	− 125	− 200

Sixth-Grade Math Minutes © 2007 Creative Teaching Press

MINUTE 18

1. Which of these has more days?

    a. 1 month   **b.** 3 weeks   **c.** 20 days

2. All of these shapes have a right angle except:

    a. ▢   **b.** ▭   **c.** ◺   **d.** ▱

3. Put these numbers in order from greatest to least: 5.06, 5.60, 0.056, 0.56.

    _____

4. Circle all fractions that are equal to $\frac{1}{3}$: $\frac{2}{6}$ $\frac{2}{5}$ $\frac{3}{9}$ $\frac{3}{8}$

5. If the pattern continues, should the last box have a dot in it?   Circle: Yes or No

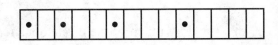

6. Which shape has a greater area? _____

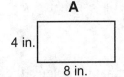

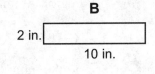

7. These five cubes were placed in a bag. What is the probability that a dark one would be pulled out of the bag first? _____

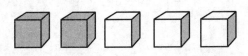

8. ▢ ÷ 4 = 13

9.   12 + 6 + 8 =   11 + 9 + 5 =   7 + 9 + 13 =

10.   15 − 4 − 6 =   21 − 10 − 2 =   20 − 6 − 3 =

Sixth-Grade Math Minutes © 2007 Creative Teaching Press

MINUTE 19

1. About how many inches long is this line segment? •————————————•

    a. 1   **b.** 3   **c.** 12   **d.** 25

2. Cross out the three-dimensional shape.

3. If $\dfrac{1}{2} \times \dfrac{3}{5} = \dfrac{3}{10}$, then $\dfrac{1}{3} \times \dfrac{4}{5} =$ _____.

For Problems 4–5, use the circle graph to the right.

4. How much of the circle does region C represent? _____

5. Is region A more or less than $\dfrac{1}{4}$? _____

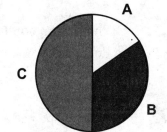

6. Find the number that completes the problem.

    2 ⬜ × 7 = 168

7. If $a = 4$, then $10a =$ _____.

8. If you rearrange the numbers of the year 2007, what is the largest number you can make? _____

9. (9)(7) =   (25)(6)=   (3)(12) =

10. $\dfrac{49}{7} =$ $\dfrac{56}{8} =$ $\dfrac{27}{9} =$

Sixth-Grade Math Minutes © 2007 Creative Teaching Press

MINUTE 20

1. Which of these has more minutes?

    a. 2 hours   **b.** 200 minutes

2. If you fit these two shapes together, which shape will you have? _____

    a.   **b.**

3. $\dfrac{2}{5} \times \dfrac{3}{7} =$

For Problems 4–6, use the Venn diagram to the right.

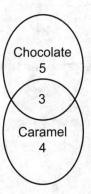

Chocolate 5

3

Caramel 4

4. How many people like chocolate only? _____

5. How many people like caramel only? _____

6. How many people like both? _____

7. If $3x = 21$, then $x =$ _____.

8. Complete the pattern.   A C E G ____.

9.
    14.3   15.8   23.4
    $-\,6.8$ $-\,4.6$ $-\,0.5$

10. $2 \cdot 3 \cdot 5 =$ $2 \cdot 2 \cdot 3 =$ $2 \cdot 5 \cdot 7 =$

Sixth-Grade Math Minutes © 2007 Creative Teaching Press

MINUTE 21

1. A state lottery might give out ten _____ dollars as a top prize.
   a. million   **b.** billion   **c.** trillion

2. Which of the following shapes has only two right angles?
   a.   **b.**   **c.**

3. $\frac{1}{2}$ of 40 =

4. $\frac{1}{3} \times \frac{1}{8} =$

5. $\frac{5+3+4}{6} =$

6. Describe the rule for this pattern: 2, 7, 6, 11, 10, 15. . . . _____

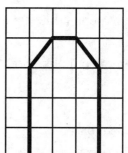

7. Find the area of the hexagon. _____

8. $2 \cdot 3 \cdot \boxed{} = 30$

9.   6,000
   $\underline{-\ 5,386}$

10.   4,508
    $\underline{-\ 1,207}$

Sixth-Grade Math Minutes © 2007 Creative Teaching Press

MINUTE 22

1. If it is 5:12 now, what time was it 24 minutes ago? _____

2. Which of the following letters has one line of symmetry? **E   F   N**

_____

3. $\frac{1}{3}$ of 9 =

4. $\frac{1}{5} \cdot \frac{4}{7} =$

5. 4(5 + 11) =

6. The third number in each of these rows is found by _____.

1	1	2
2	3	5
5	10	15
10	10	20

7. Find the perimeter of the shape to the right. _____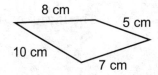

8. Find the sum of the second (shaded) column. _____

1	2	9
5	8	6
4	3	7

9. 16 ÷ 4 =   18 ÷ 3 =   15 ÷ 5 =

10. 34   56
      × 3   × 4

Sixth-Grade Math Minutes © 2007 Creative Teaching Press

MINUTE 23

1. Round each number to the nearest 1,000.

1,238 = _____   1,850 = _____   3,320 = _____

2. Which of the following letters has two lines of symmetry?   **H   W   L   V**

_____

3. $\frac{1}{4}$ of 12 =

4. If $\frac{1}{5} + \frac{1}{5} = \frac{a}{5}$, then $a =$ _____.

5. $3 + 4 \cdot 2 + 6 =$

6. Complete the pattern box.

2	5	8	12
10	25		

7. If the perimeter of this shape is 25, then $x =$ _____.

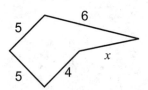

8. The sum of the third (shaded) column is _____.

1	2	9
5	8	6
4	3	7

9. $9 \times 6 =$ $9 \cdot 8 =$ $(9)(9) =$

10. ☐ $\div 4 = 9$ ☐ $\div 6 = 8$ ☐ $\div 5 = 7$

Sixth-Grade Math Minutes © 2007 Creative Teaching Press

MINUTE 24

1. What is the value in cents of 10 quarters and 2 dimes? _____

2. Which of the following represents a line?

   a.   **b.**   **c.**

3. Which fraction represents $15 \div 2$?

   a. $\dfrac{2}{15}$ **b.** $\dfrac{15}{2}$ **c.** $\dfrac{15}{15}$ **d.** $\dfrac{2}{2}$

4. $\dfrac{2}{7} + \dfrac{3}{7} =$

5. $4 + 7 + \boxed{} = 32$

6. Fill in the empty square to the right by following the pattern given.

3	8		6
9	24	30	18

7. The width of a rectangle is 4 feet. If the area is 36 ft.^{2}, then the length = _____.

8. Find the sum of the first column. _____

1	2	9
5	8	6
4	3	7

9.   86   93
       $\times\,10$ $\times\,10$

10.   50   60
       $\times\,50$ $\times\,60$

Sixth-Grade Math Minutes © 2007 Creative Teaching Press

MINUTE 25

1. Kelly has $10, which is $2 more than Tina has. How much money does Tina have?
_____

2. Which of the following represents a ray?

    a.   **b.**   **c.**

3. Which of the following represents the division problem $16 \div 9$ as a fraction?

    a. $\dfrac{9}{16}$ **b.** $\dfrac{16}{16}$ **c.** $\dfrac{16}{9}$ **d.** $\dfrac{6}{19}$

4. $\dfrac{5}{7} + \dfrac{6}{7} =$

5. Use $+$, $-$, $\times$, or $\div$ to complete. $7\ \boxed{}\ 5 = 35$

6. How many sides should the next shape in the pattern have? _____

7. If every side of an octagon is 6 inches, what is the perimeter? _____

8. What is the product of the first (shaded) row? _____

1	2	9
5	8	6
4	3	7

9. Find the remainders for $3\overline{)14}$ and $5\overline{)17}$. _____

10. $\dfrac{1}{2}$ of 12 = $\dfrac{1}{2}$ of 18 =

Sixth-Grade Math Minutes © 2007 Creative Teaching Press

MINUTE 26

1. Bobby is 7 years old. Ray is twice Bobby's age. How old is Ray? _____

2. Which of the following represents a line segment?

   a. ←———→   **b.** •———•   **c.** •———→

3. All of the following mean 21 divided by 9 except:

   a. $\dfrac{21}{9}$ **b.** $\dfrac{9}{21}$ **c.** $21 \div 9$ **d.** $9\overline{)21}$

4. If $4\dfrac{1}{2} = \dfrac{9}{2}$, then $6\dfrac{1}{2} =$ _____.

5. $5 \times (8 + 2) =$

6. Complete the pattern: A B A A B A A A B A A A ____.

7. Find the area of the triangle. _____

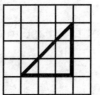

8. Find the product of the numbers in the third row. _____

1	3	9
5	8	6
4	2	7

9. $7\overline{)420} =$ $3\overline{)1,500} =$

10.
```
  8,359
+ 6,728
```

MINUTE 27

1. Describe how you could have $0.87 with the least number of coins possible.

_____

2. Which circle has a radius drawn on it?

a.   b.   c.

3. $\dfrac{8}{12} + \dfrac{3}{12} =$

4. If $5\dfrac{1}{3} = \dfrac{x}{3}$, then $x =$ _____.

5. $(5 \times 6) + (3 \times \boxed{}) = 36$

6. Complete the pattern. 64, 32, 16, 8, _____, _____.

7. What is the area of the shape to the right? _____

8. How many eggs did Lucky lay? _____

Eggs Laid Last Season

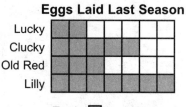

Each = 1 dozen

9.   9,476
   − 1,355

10.   2,761
   + 3,478

Sixth-Grade Math Minutes © 2007 Creative Teaching Press

NAME: _____

MINUTE 28

1. Sarah has three dozen cookies. How many cookies is this? _____

2. The hypotenuse is the longest side of a right triangle. Which letter is beside the hypotenuse in this triangle? _____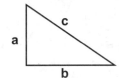

3. $3\frac{1}{3} + 2\frac{1}{3} =$

4. If $\frac{n}{6} = \frac{1}{2}$, then $n =$ _____.

5. Use + or × to complete the problem. $\frac{1}{5} \; \square \; \frac{2}{5} = \frac{3}{5}$.

6. Which of the following triangles would be the next in this pattern?

   a.   b.   c.

7. Find the product of the third column. _____

9	3	1
5	8	6
4	2	7

For Problems 8–9, use the bar graph to the right.

8. According to the graph, how many students are getting an A for their math grade? _____

9. According to the graph, are there more Bs or Ds in the class? _____

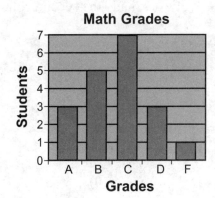

10. $6 \times 0.2 =$ _____ $7 \times 0.4 =$ _____ $8 \times 0.5 =$ _____

MINUTE 29

1. If today is Monday, what day will it be eight days from now? _____

2. Which letter is beside the hypotenuse? _____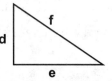

3. Write as an improper fraction: $8\dfrac{1}{3} =$

4. Use >, <, or = to complete the problem. $0.55\ \boxed{}\ \dfrac{1}{2}$

5. Use + or × to complete the problem. $\dfrac{1}{5}\ \boxed{}\ \dfrac{2}{5} = \dfrac{2}{25}$

6. Complete the sequence: Z 1 Y 2 X 3 ____ .

7. Find the area of the shape. _____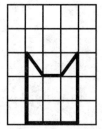

8. Shade the second column of the grid used in Problem 7.

9. $1.3 + 0.2 + 0.4 =$ $0.8 + 0.2 + 0.7 =$

10. $(2)(4)(5) =$
    $(4)(5)(1) =$
    $(5)(6)(0) =$

Sixth-Grade Math Minutes © 2007 Creative Teaching Press

MINUTE 30

1. If tomorrow is the 4th of June, what day will it be three days from today?

_____

2. The following three numbers are the side lengths of a right triangle: 5, 12, and 13. Which number is the length of the hypotenuse? _____

3. $\dfrac{4}{9} \cdot \dfrac{1}{3} =$

For Problems 4–5, use the grid to the right.

4. The grid has 100 boxes.
How many of them are shaded? _____

5. How many boxes in the grid are not shaded? _____

6. $4(3 + 9) =$

7. What is the perimeter of the shape to the right? _____

8. Complete the chart:

x	8	10	16
y	4	5	

9. Use >, <, or = to complete the problem. $0.75 \boxed{} \dfrac{3}{4}$

10. $\dfrac{1}{5} \cdot \dfrac{1}{4} =$ $\dfrac{1}{7} \cdot \dfrac{2}{3} =$ $\dfrac{1}{10} \cdot \dfrac{3}{4} =$

MINUTE 31

1. If today is Tuesday, what day will it be three weeks from tomorrow? _____

2. Which circle has a diameter drawn on it?

a.   b.   c. [circle]

For Problems 3–4, use the grid to the right.

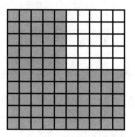

3. How many boxes in the grid are shaded? _____

4. What fraction of the grid is shaded? _____

5. Which would have the greater perimeter, the circle or the box? _____

6. $2 \cdot 2 \cdot 2 \cdot \boxed{} = 40$

7. Complete the sequence: 0, 5, 1, 6, 2, 7, _____, _____.

8. The following cubes are placed into a bag. What is the probability that a cube with the letter B will be drawn from the bag? _____

9. $3.65 \times 100 =$ $2.7 \times 100 =$

10. $4\overline{)1,236} =$ $5\overline{)1,235} =$

MINUTE 32

1. Ted gets paid every two weeks. Is it possible for Ted to get three paychecks in one month?   Circle: Yes   or   No

2. Which circle has a chord drawn on it that is not a diameter?

a.   b.   c.

3. $\dfrac{8}{5} - \dfrac{3}{5} - \dfrac{2}{5} =$

For Problems 4–5, use the grid to the right.

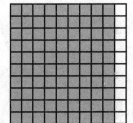

4. What percent of the grid is shaded? _____

5. What percent of the grid is not shaded? _____

6. Which would have the greater area, the circle or the box? _____

For Problems 7–9, use the chart to the right.

7. $a + b =$

8. $b \cdot d =$

9. $\dfrac{e}{b} =$

Letter	Value
a	2
b	4
c	0
d	5
e	8

10. $\dfrac{1}{5} + \dfrac{2}{5} =$ $\dfrac{1}{5} \cdot \dfrac{2}{5} =$

MINUTE 33

For Problems 1–3, circle the amount that is larger.

1. 2 months   or   10 weeks

2. 9 quarters   or   2 dollars

3. 4 feet   or   50 inches

For Problems 4–5, use the diagram to the right.

4. Triangle 1 is _____ of the square.

    a. $\dfrac{1}{2}$ **b.** $\dfrac{1}{3}$ **c.** $\dfrac{1}{4}$

5. Triangle 2 is _____ of the square.

    a. $\dfrac{1}{2}$ **b.** $\dfrac{1}{3}$ **c.** $\dfrac{1}{4}$

6. Sally says the perimeter of the rectangle is 22. Desiree says that it is 18. Who is correct? _____

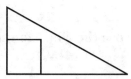

7. Which shape has the greater area—the triangle or the square? _____

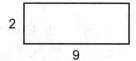

For Problems 8–10, use the chart to the right.

8. $d - a =$

9. $\dfrac{c + 8}{5} =$

10. If the pattern continues, what would be the value of the letter e? _____

Letter	Value
a	24
b	28
c	32
d	36
e	

Sixth-Grade Math Minutes © 2007 Creative Teaching Press

MINUTE 34

1. The number 1 with three zeros after it would represent _____.

    a. one thousand   **b.** ten thousand   **c.** one million

2. Match the name of each word with its figure.

    rhombus   **a.**

    square   **b.**

    quadrilateral   **c.**

For Problems 3–4, use the grid to the right.

3. What percent of the grid is shaded? _____

4. If 50% of the grid is supposed to be shaded, how many more boxes would need to be shaded? _____

5. $16 \div 2 \div 2 \div 2 =$

6. Fill in the box with the next number in the sequence:

     5,394,600
     5,494,600
     5,594,600

7. An electric fence around a property would be most like the _____ of the property.

    a. area   **b.** volume   **c.** perimeter

8. $4 \cdot 5 = \boxed{} + 5$

9. $2.36 \times 10 =$ $0.34 \times 100 =$ $46 \times 10 =$

10. $\frac{1}{2}$ of 40 = $\frac{1}{2}$ of 50 =

Sixth-Grade Math Minutes © 2007 Creative Teaching Press

MINUTE 35

1. The number 435 should be written as:

    a. four hundred and thirty five   **b.** four hundred thirty five

    c. four hundred thirty-five

2. Which set of shapes shows two figures that are NOT congruent (same size and shape)?

  a.   **b.**   **c.**   **d.**

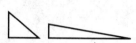

For Problems 3–5, circle the larger amount.

3. $\dfrac{1}{4}$ or   75%

4.   25%   or $\dfrac{1}{2}$

5. $\dfrac{9}{10}$ or   95%

6. The following cards were numbered as shown, placed face down on a table, and then mixed up. If a card is turned over randomly, what number would show up most often? _____

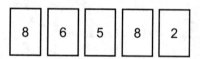

7. $4 \times 9 = \boxed{} \times 6$

8. Which of the following graphs shows these ages the best? Bob = 12, Tim = 11, Vern = 6.

  a.   **b.**   **c.**

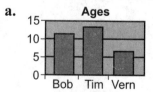
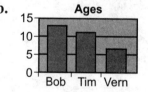
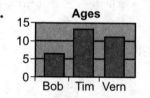

9. $14 \times 2 =$ $21 \times 2 =$ $30 \times 2 =$

10. $\dfrac{1}{3} \times \dfrac{1}{3} =$ $\dfrac{1}{3} + \dfrac{1}{3} =$ $\dfrac{1}{3} - \dfrac{1}{3} =$

Sixth-Grade Math Minutes © 2007 Creative Teaching Press

MINUTE 36

1. About how wide is this paper?

    a. 9 centimeters   **b.** 9 inches   **c.** 9 feet

2. How many lines of symmetry does this shape have? _____

3. $\dfrac{1}{8} + \dfrac{3}{8} + \dfrac{3}{8} =$

4. What fraction does the letter B represent? _____

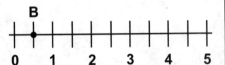

For Problems 5–7, cross out the number that does NOT belong on the list with the others.

5.   2   8   10   13

6.   7   4   11   19

7.   6   15   25   35

8. The four cards below form a pattern. What would the 7th card look like?

1	2	3	4
5	10	15	20

7th Card =

9.

$$\begin{array}{r} \square \\ +\ 14 \\ \hline 22 \end{array}$$

10. Complete the table by finding the sum and product of the two numbers in each row.

Number	Number	Sum	Product
5	8		
7	9		

NAME: _____

MINUTE 37

1. Each song on Mel's MP3 player is most likely to be about _____ long.
    a. 3 minutes   **b.** 3 seconds   **c.** 30 minutes   **d.** 3 hours

2. Match each word with its figure.
    Line   **a.** •————•
    Segment   **b.** ←————→
    Ray   **c.** •————→

3. What fraction does the letter A represent? _____

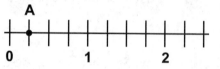

4. Which number should go in the box?   340, 344, ☐ , 352
    a. 345   **b.** 346   **c.** 352   **d.** 348

5. $200{,}000 + 50{,}000 + 8{,}000 + 100 + 4 =$ _____
    a. 205,841   **b.** 258,140   **c.** 258,104   **d.** 250,814

6. I am an odd number between 10 and 20. I can be divided by 3. What number am I? _____

7. As Martha wrapped a present in wrapping paper, she happened to think that the paper was most like the _____ of the box.
    a. surface area   **b.** perimeter   **c.** volume

For Problems 8–9, use the graph to the right.

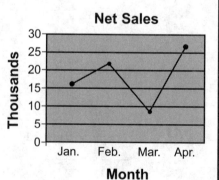

8. According to the graph, sales in March were _____.
    a. up   **b.** down
    c. about the same as the other months

9. In which month were the sales the best? _____

10. $56.2 \div 10 =$ $426 \div 10 =$ $5.8 \div 10 =$

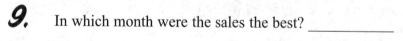

Sixth-Grade Math Minutes © 2007 Creative Teaching Press

NAME:

MINUTE 38

1. Which of the following might be reasonable dimensions of a bathroom?

    a. 40 in. × 30 in.   **b.** 2 miles × 3 miles   **c.** 9 feet × 12 feet

2. Which set of shapes show two figures that are congruent?

  a. 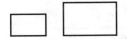   **b.**

  c.   **d.**

3. What fraction should be next in this sequence? $\frac{1}{12}, \frac{3}{12}, \frac{5}{12},$ _____.

4. $\frac{1}{2} \times \frac{3}{4} \times \frac{5}{6} =$

For Problems 5–6, use the grid to the right.

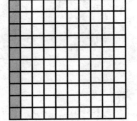

5. What percent of the grid is shaded? _____

6. How many more squares would have to be filled in so that half of the grid is shaded? _____

7. I am an even number between 1 and 10. I can be divided by 3 evenly. What number am I? _____

8. Justine says that the area of the rectangle is 24 square units. Marcie says that it is 20 square units. Who is correct?

_____

4   6

9. 7 • 7 =   8 × 8 =   (6)(6) =

10. 7 ÷ 0.7 =   8 ÷ 0.8 =   6 ÷ 0.6 =

Sixth-Grade Math Minutes © 2007 Creative Teaching Press

MINUTE 39

1. $400{,}000 + 5{,}000 + 800 + 6 =$
    a. 450,860   **b.** 450,806   **c.** 405,860   **d.** 405,806

2. How many sides does each shape have?
    Pentagon: _____ Octagon: _____ Decagon: _____

3. If 45% = 0.45, then 55% = _____.

4. Sales tax in a particular state might be: **a.** 6%   **b.** 60%

5. What number is twice as much as 2,400?
    a. 4,800   **b.** 1,200
    c. 4,400   **d.** 480

6. What is the error in the problem $32 \times 9 = 281$? _____

7. Sally believes that the perimeter of this rectangle is 32.
    What mistake did she make? _____

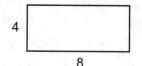
4

8

For Problems 8–10, use the chart to the right.

8. $a + b =$

9. $b \times c =$

10. $\dfrac{(a)(c)}{2} =$

a	b	c
5	4	6

Sixth-Grade Math Minutes © 2007 Creative Teaching Press

MINUTE 40

1. The number 1 with six zeros after it would represent _____.

    a. one thousand   **b.** ten thousand   **c.** one million

2. First do the addition; then reduce the fraction: $\dfrac{3+9+2}{8+10+2} =$

3. Write as a decimal: 61% =

4. Write as a percent: 0.47 =

5. If you add 5 to the product of 4 and 6, you get _____.

6. Complete the pattern by filling in the bottom box.

5	+	4
	9	
2	+	5
	7	
1	+	4

7. $3 + 2 \times \boxed{} = 17$

For Problems 8–10, round to the underlined place value.

8. 3<u>3</u>.28 _____

9. 0.0<u>5</u>61 _____

10. <u>3</u>47.5 _____

Sixth-Grade Math Minutes © 2007 Creative Teaching Press

MINUTE 41

1. If you rearrange the numbers in the year 1942, what is the smallest number you can make? _____

For Problems 2–3, use the diagram to the right.

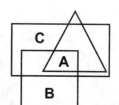

2. Which letter is outside the rectangle but inside the square? _____

3. Which letter is inside all three shapes? _____

4. Complete the chart.

Fraction	Decimal	Percent
$\dfrac{3}{4}$		
	0.1	

For Problems 5–6, use the Venn diagram to the right.

Favorite Kinds of Movies

Scary 10 — 3 — Funny 8

5. How many people prefer scary movies only (not funny)? _____

6. How many people took part in this survey? _____

7. The *y* numbers in this chart are _____ times the *x* numbers.

x	3	8	12
y	12	32	48

For Problems 8–10, estimate to find the best answer.

8. 22 + 51 is approximately
    a. 70   **b.** 80   **c.** 60   **d.** 100

9. 96 + 103 is approximately
    a. 100   **b.** 300   **c.** 200   **d.** 400

10. 21 × 29 is approximately
    a. 500   **b.** 400   **c.** 300   **d.** 600

Sixth-Grade Math Minutes © 2007 Creative Teaching Press

MINUTE 42

1. The number 1 with four zeros after it would represent _____.
  a. one thousand **b.** ten thousand **c.** one million

For Problems 2–3, use the diagram to the right.

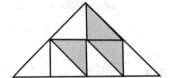

2. What fraction of the large triangle is shaded? _____

3. If one more triangle were shaded, what percent of the
large triangle would be shaded? _____

4. Complete the chart.

Fraction	Decimal	Percent
		25%
	0.3	

5. One more than the product of 8 and 10 is _____.

6. Which of the following is NOT a multiple of 6? _____   6   12   24   32   36

7. Which of the following would be the correct order for simplifying a math problem?
  a. exponents, parentheses, multiplying, adding
  b. multiplying, dividing, subtracting, parentheses
  c. parentheses, adding, multiplying, exponents
  d. parentheses, exponents, multiplying, adding

8. Complete the pattern: 1, 3, 6, 8, 11, 13, 16, _____.

9. $\frac{1}{8} + \frac{3}{8} =$ $\frac{1}{8} \times \frac{3}{8} =$

10. $20 \cdot 10 =$ $30 \cdot 5 =$ $40 \cdot 2 =$

Sixth-Grade Math Minutes © 2007 Creative Teaching Press

49

MINUTE 43

1. The correct way to write 12.36 would be:
   a. Twelve and thirty-six hundredths
   b. Twelve and thirty-six thousandths
   c. Twelve and thirty six tenths
   d. Twelve and thirty and six hundredths

2. Put these three angles in order from least to greatest: Right, Obtuse, Acute.

   _____

3. Which two letters represent the hypotenuse of a triangle
   in this figure?
   a. AD   b. AB
   c. BC   d. AC

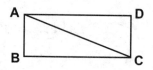

For Problems 4–6, cross out the item that does NOT belong in each list.

4. 4   8   12   15

5. fence   walls   frame   carpet

6. days   inches   weeks   months

7. If $7 \times 6 = 5 \times 8 + x$, then $x =$ _____.

For Problems 8–10, use the chart to the right.

8. $ab =$ $ac =$

9. $\dfrac{c}{a} =$ $\dfrac{c}{b} =$

10. $a + b + c =$

a	b	c
3	5	30

Sixth-Grade Math Minutes © 2007 Creative Teaching Press

MINUTE 44

1. The 1984 Olympics were in Los Angeles. If the Olympics occur every four years, which of these years did not have an Olympics?

    a. 1988   **b.** 1996   **c.** 2002   **d.** 2004

2. Each side of the cube is called a face.
How many faces does a cube have? _____

3. If $3.\overline{8}$ means 3.88888888…, then how would you write 1.77777777…? _____

4. $20\% + 30\% =$

5. If $\dfrac{1}{2} \times 10 = 8 - x$, then $x =$ _____.

For Problems 6–8, estimate to find the best answer. (**Hint:** "≈" means "approximately")

6. $26 + 73 + 41 \approx$

    a. 120   **b.** 140   **c.** 160

7. $\$1.78 + \$2.99 + \$0.84 \approx$

    a. $6   **b.** $3   **c.** $4

8. $8 + 11 + 12 + 17 \approx$

    a. 30   **b.** 40   **c.** 50

9. $\dfrac{7}{9} - \dfrac{4}{9} =$ $\dfrac{7}{9} + \dfrac{4}{9} =$

10. Change to improper form: $9\dfrac{1}{2} =$ $10\dfrac{1}{4} =$

MINUTE 45

1. How many 50-cent cans of soda can be purchased with $5? _____

2. How many faces does this shape have? _____

3. What is another way to write 0.8222222222…? _____

4. Rewrite in decimal form: $\dfrac{25}{100} =$

5. If $12 + m = 22$, then $m =$ _____.

6. Complete the pattern. A B B C C C ____ ____ ____ ____.

For Problems 7–8, use the chart to the right.

7. Which grade had about the same number of honor roll students both years? _____

8. Which grade should be most concerned about the trend from 2006 to 2007? _____

Honor Roll Students		
Grade	2006	2007
3	51	83
4	46	47
5	90	46

9.
$\begin{array}{r} 0.952 \\ -\,0.841 \\ \hline \end{array}$
$\qquad$
$\begin{array}{r} 0.855 \\ -\,0.704 \\ \hline \end{array}$

10. $\dfrac{100}{10} =$ $\qquad$ $\dfrac{1,000}{10} =$ $\qquad$ $\dfrac{10,000}{10} =$

Sixth-Grade Math Minutes © 2007 Creative Teaching Press

MINUTE 46

1. Using the numbers 1–6, fill in the blanks to create the largest number possible:

____ ____ ____ ____ . ____ ____

2. If the dotted line is a line of symmetry, how long is side AB? _____

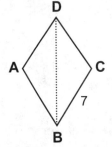

3. Write using bar notation: 0.39393939… =

4. If $\dfrac{x}{100} = 0.3$, then $x =$ _____.

5. $10\% + 25\% + 20\% =$

For Problems 6–7, use the chart to the right.

6. Complete the chart.

x	y
2	10
3	15
5	

7. If $y = 35$, then $x =$ _____.

8. Use the graph to find the values of A, B, and C.

A = _____   B = _____   C = _____

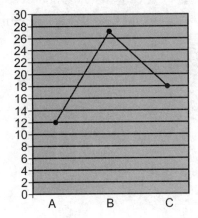

9. $\dfrac{1}{10} + \dfrac{1}{10} =$ $\dfrac{1}{10} \cdot \dfrac{1}{10} =$

10. $(0.5)(0.6) =$ $(0.4)(0.7) =$

MINUTE 47

1. Using the numbers 1–6, fill in the blanks to create the smallest number possible:

_____ _____ _____ _____ . _____ _____

2. Fill in the missing number in the factor tree.

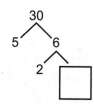

3. Is 34.82 closer to 34 or 35? _____

4. Eight hundredths plus nine hundredths equals _____.

5. $\frac{1}{2} + 0.2 =$

6. 20.4% + 20.5% + 4.1% =

For Problems 7–8, use the circle graph to the right.

Election Survey

7. What percent of the votes did candidate A receive? _____

8. If candidates A and C combined their votes, they would have _____ candidate B.

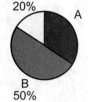

a. more than   **b.** less than   **c.** the same as

9. 0.98 × 10 =   0.98 × 100 =   0.98 × 1,000 =

10. $\frac{5}{100} = $ _____% $\frac{15}{100} = $ _____% $\frac{85}{100} = $ _____%

Sixth-Grade Math Minutes © 2007 Creative Teaching Press

MINUTE 48

1. Mike can ride his bike 15 miles per hour. How many miles could he reasonably ride in one day?

a. 350   **b.** 250   **c.** 500   **d.** 100

2. Would five smaller boxes fit inside the larger box?
Circle: Yes   or   No   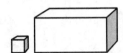

3. $0.4 + 0.3 + 0.2 =$

4. Circle the greater value: $0.\overline{7}$ or 0.7

5. $\frac{3}{4} + 0.20 + 0.05 =$

For Problems 6–8, round to the place value of the underlined number.

6. $0.\underline{6}15$ _____

7. $\underline{9}3$ _____

8. $10\underline{5}.87$ _____

9. Find the areas of the rectangles described in the chart.
_____ _____ _____

rectangle	length	width	area
1	6	7	
2	9	10	
3	5	9	

10. Circle the problems below that have whole number answers (not decimal or fractional answers).

$24 \div 5$ $\frac{200}{10}$ 0.16×100 $\frac{1}{4} + \frac{1}{4} + \frac{1}{4}$

MINUTE 49

1. It took Jill two hours to drive 100 miles. What was her average speed? _____

2. Would 10 smaller cans fit inside the larger can?
Circle: Yes   or   No

3. Circle the greater value: $0.\overline{7}$ or $\dfrac{3}{4}$

4. Complete the chart.

Fraction	Decimal	Percent
		40%
$\dfrac{1}{4}$		

5. If $x = y$ and $y = 2$, then $3x =$ _____.

6. Complete the chart.
(**Hint:** The product of each column equals the same value.)

1	2	3	4
24	12	8	

7. James has $2.85. Fill in the remaining box to show how many nickels he has.

Quarters	Dimes	Nickels
8	7	

8. Complete the pattern.   AC BD CE DF _____.

9. Circle the following fractions that are equal to $\dfrac{1}{5}$. $\dfrac{2}{10}$ $\dfrac{4}{20}$ $\dfrac{5}{20}$ $\dfrac{10}{40}$

10. How many days are in each of the following?
2 weeks = _____
1 year (not a leap year) = _____
3 days more than 5 weeks = _____

MINUTE 50

1. A case holds four boxes. A box holds five cartons. How many cartons are in two cases? _____

2. How many faces does this shape have? _____

For Problems 3–5, use < , >, or = to complete.

3. 8.13 _____ 8.4

4. 0.004 _____ 0.05

5. 0.$\overline{4}$ _____ 0.4

6. Complete the factor tree.

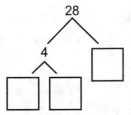

7. Can the numbers you wrote in the empty boxes in Problem 6 be divided by other numbers besides 1 and the numbers themselves?   Circle: Yes   or   No

8. How much money was raised by all the children?

_____

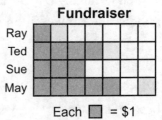

Fundraiser

Each ☐ = $1

9. 3 + ☐ = 18   3 × ☐ = 18   18 ÷ ☐ = 3

10. 40 dimes = _____ dollars
40 nickels = _____ dollars
40 quarters = _____ dollars

Sixth-Grade Math Minutes © 2007 Creative Teaching Press

MINUTE 51

1. Joanne has 15 basketball cards. Jackie has 8. If Joanne gives Jackie 5 of her cards, how many will each girl have? Joanne: _____ Jackie: _____

2. What is the total number of degrees in a triangle? _____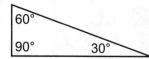

3. Write using bar notation: 0.38888888… =

4. If $\sqrt{9}$ = 3, then $\sqrt{16}$ = _____ .

5. Becky is the same height as Brittany. Brittany is the same height as Mandy. Are Becky and Mandy the same height?   Circle: Yes   or   No

6. If $3x + 2 = 11$, could $x = 5$?   Circle: Yes   or   No

For Problems 7–8, use the graph to the right.

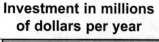

7. Which company (A, B, or C) made the poorest investment in one year? _____

8. Which company (A, B, or C) made the best investment in one year? _____

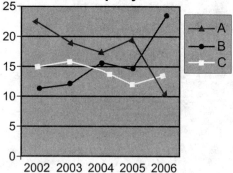

9. How many sides does each of these shapes have?
Rectangle: _____ Pentagon: _____ Octagon: _____

10. Change to an improper fraction: $5\frac{1}{3}$ = $6\frac{2}{3}$ = $3\frac{1}{4}$ =

Sixth-Grade Math Minutes © 2007 Creative Teaching Press

MINUTE 52

1. Which of the following numbers is one billion?
   a. 1,000,000   **b.** 1,000,000,000   **c.** 1,000,000,000,000

2. What is the missing angle? _____

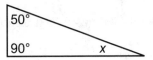

3. $0.3 + 40\% + \dfrac{1}{4} =$

4. $\sqrt{25} =$

5. The letters A, B, and C can be arranged in six ways. Five ways are listed below.
   Find the sixth way.
   ABC   ACB   BAC   BCA   CAB   _____

For Problems 6–8, solve if $a = 10$, $b = 5$, and $c = 3$.

6. $12.4 \times a =$

7. $\dfrac{a+b}{c} =$

8. $a + b \cdot c =$

9. In Problem 8, which operation should you do first?
   a. add   **b.** subtract   **c.** multiply   **d.** divide

10. What is the area of this rectangle? _____   3 mm
   What is the perimeter of this rectangle? _____

          6 mm

Sixth-Grade Math Minutes © 2007 Creative Teaching Press

MINUTE 53

1. Jason drove for three hours at an average speed of 55 miles per hour. How far did he go? _____

2. The interior angles of a triangle add up to _____ degrees.

3. Circle all of the following that are equal to $\frac{3}{10}$:   0.3   3% $\frac{6}{10}$

4. $\sqrt{(4)(9)} =$

5. Fill in the missing number.

$3 \begin{cases} 6 \longrightarrow 12 \longrightarrow \boxed{} \\ 9 \longrightarrow 27 \longrightarrow 81 \end{cases}$

6. Two times a number is 14. What is the number? _____

7. If the pattern continues, should the last box be shaded or clear? _____

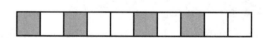

8. Allan has $3.05. Fill in the remaining box to show how many dimes he has.

Quarters	Dimes	Nickles
8		1

9.
$\begin{array}{r} 67 \\ -28 \\ \hline \end{array}$

$\begin{array}{r} 92 \\ -45 \\ \hline \end{array}$

$\begin{array}{r} 101 \\ -33 \\ \hline \end{array}$

10.   (3)(4)(3) =   (2)(5)(3) =

Sixth-Grade Math Minutes © 2007 Creative Teaching Press

MINUTE 54

1. Michaela makes $5.50 per hour at her job. How much does she make in an average eight-hour day? _____

2 What is the area of the shape? _____

3. $\frac{1}{2}(3 \cdot 4 + 4) =$

4. $\sqrt{36} \cdot \sqrt{81} =$

5. If a coin were tossed on the grid in Problem 2, would it have a better chance of landing inside or outside of the shape? _____

6 Five more than five times a number is 30. What is the number? _____

7. Place ( ) symbols in this problem to make a true statement. $3 + 9 \times 4 = 48$

8. Point B is two units larger than point A. What number represents the value of point B?_____

9. Draw the next B in the pattern.

B ⅊ B

10. Find the sum of each row.

	Sum		
_____	3	5	7
_____	6	8	1
_____	9	4	2

MINUTE 55

1. 5,649 rounded to the nearest: 10 = _____   1,000 = _____

2. If both the length and width of this rectangle are doubled, what will the new area be? _____

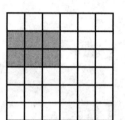

3. Circle the two smallest numbers.

3.68   3.06   3.7   3.08   36.8   3.068

4. If $7^2 = 7 \times 7 = 49$, then $8^2 =$ _____.

For Problems 5–6, use the spinners to the right.

5. How many possibilities could occur if both spinners are spun? _____

6. What is the probability of getting an A and then a 2? _____

7. Fill in the missing prime numbers between 2 and 30.

2	3		7	11	13		19	23	

For Problems 8–10 use > , < , or =.

8. $\sqrt{100}$ _____ $\dfrac{20}{2}$

9. 2.8 _____ $2.\overline{7}$

10. $\dfrac{2}{3}$ _____ $\dfrac{1}{2}$

Sixth-Grade Math Minutes © 2007 Creative Teaching Press

MINUTE 56

For Problems 1–2, use the calendar to the right.

MAY

S	M	T	W	T	F	S
				1	2	3
4	5	6	7	8	9	10
11	12	13	14	15	16	17
18	19	20	21	22	23	24
25	26	27	28	29	30	31

1. Sixteen days after May 4 would be a:
a. Monday   **b.** Tuesday
c. Wednesday   **d.** Thursday

2. Which Tuesday has a date that is a prime number? _____

3. Match the letters to the numbers using the number line.

A   B   C

2   2.5   3

_____ = 2.4
_____ = 2.8
_____ = 2.1

4. Cross out any prime numbers from the grid.

5	8	12	15	21	23

	Boys	Girls	Total
Class 1	10	15	25
Class 2	18	12	30

5. What is the probability that a student pulled at random from Class 1 is a boy? _____

6. What would the next shape in this pattern be? ☐ ✚ ◯ ◯ ✚ ☐ ☐ ✚

a. ☐   **b.** ◯   **c.** ✚

For Problems 7–10, match each description with its correct mathematical expression.

7. Twice a number.   **a.** $\dfrac{n}{2}$

8. A number to the second power.   **b.** $\sqrt{n}$

9. A number divided by 2.   **c.** $2n$

10. The square root of a number.   **d.** n^2

MINUTE 57

1. Monique weighs 84 pounds. When she is holding her baby brother, she weighs 96 pounds. How much does Monique's baby brother weigh? _____

2. Complete the factor tree.

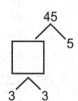

3. What is the common denominator for $\frac{1}{3} + \frac{1}{2}$? _____

4. $\frac{1}{2}(3 + 5)^2 =$

5. What is the probability that a student pulled at random from Class 1 is a girl? _____

	Boys	Girls	Total
Class 1	10	15	25
Class 2	18	12	30

6. Which one of the following solves this problem? $2x + 3 = 15$
    a. $x = 5$ **b.** $x = 4$ **c.** $x = 7$ **d.** $x = 6$

7. Complete the analogy: ◯ is to ⊖ as ☐ is to:

    a. ⊖   **b.** ▭   **c.** ▭

8. Find two pairs of different (unequal) odd numbers that complete the equation.

    ☐ + ☐ = 10

9. Fill in the missing numbers to complete the chart.

Numbers	Sum	Difference	Product
1,4	5		4
2,8	10	6	

10. If $x^2 = 16$, then $x =$ _____.

Sixth-Grade Math Minutes © 2007 Creative Teaching Press

MINUTE 58

1. A new car is available in five different colors and with two different types of engines. How many different combinations of colors and engines could you order? _____

2. The perimeter of the rectangle to the right is 24 ft. What is the width? _____

x ☐ 7 ft.

3. What would the area in Problem 2 be? _____ (**Hint:** Use the width you found.)

4. $(3 + 7)^2 =$

5. What is the total number of boys in these classes? _____

Quarters	Boys	Girls	Total
Class 1	10	15	25
Class 2	18	12	30

For Problems 6–7, use the Venn diagram to the right.

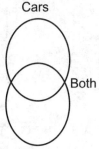

6. Fill in the Venn diagram using the following information.
Four people only drive cars.
Six people only drive trucks.
Eight people drive both cars and trucks.

7. How many people took part in the survey in Problem 6? _____

8. Fill in the missing numbers. $3 \times 4 =$ ☐ $\div 6 =$ ☐

9.
5,122
+ 2,308

10. $\dfrac{1}{5} + \dfrac{3}{5} =$ $\qquad$ $\dfrac{1}{5} \cdot \dfrac{3}{5} =$

Sixth-Grade Math Minutes © 2007 Creative Teaching Press

MINUTE 59

For Problems 1–2, use the calendar to the right.

S	M	T	W	T	F	S
				1	2	3
4	5	6	7	8	9	10
11	12	13	14	15	16	17
18	19	20	21	22	23	24
25	26	27	28	29	30	31

MAY

1. Three weeks later than Friday, May 2, would be Friday, May _____.

2. Sandy gets paid every Friday. How many paychecks will she get in the month of May? _____

3. If each box is two units long, find the perimeter of the shaded rectangle. _____.

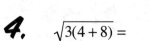

4. $\sqrt{3(4+8)} =$

5. Find the next card in the pattern.

2	4	6	8
5	8	11	14

5th Card =

6. Which of these numbers should go inside the box to make the equation true?

$$\frac{\boxed{} + 4}{2} = 10$$

a. 12   **b.** 20   **c.** 16

7. If 10% of this grid were shaded, how many squares would be shaded? _____.

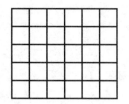

For Problems 8–10, evaluate if $a = 2$, $b = 4$, and $c = 12$.

8. The sum of a and $c =$

9. $6^a =$

10. $\dfrac{c}{3b} =$

Sixth-Grade Math Minutes © 2007 Creative Teaching Press

MINUTE 60

1. What is the best estimate of the time on this clock?
a. 3:55 **b.** 4:55 **c.** 2:55 **d.** 3:15

2. Fill in the square with the correct fraction. $12 \times \boxed{} = 6$

3. Which shape is congruent to this one?

a.   **b.**   **c.**

4. $3^2 + 2^2 =$

5. What is the total number of girls in these classes? _____

	Boys	Girls	Total
Class 1	10	15	25
Class 2	18	12	30

6. If each of these hearts could be colored red, pink, or blue, how many different ways could they be colored? (**Hint:** More than one heart could be the same color.)

_____

For Problems 7–10, match each expression with its correct description.

7. $\dfrac{n}{3}$ **a.** A number to the third power.

8. $3n$ **b.** Three times a number.

9. n^3 **c.** The sum of a number and three.

10. $n + 3$ **d.** A number divided by three.

MINUTE 61

1. Round each number to the underlined position.

1<u>2</u>8 =   <u>3</u>,158 =   488.3<u>7</u> =

2. How many cubes are in this shape? _____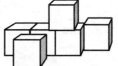

x	y
0.2	0.8
0.3	1.2
0.5	2
0.7	2.8

3. The numbers in the y column are _____ times bigger than those in the x column.

4. What number solves this equation? ☐ $\times (3 + 8) = 55$

5. Fifty tickets were sold for the lottery. Jackson bought five tickets. What are the chances he will win? _____

6 Fill in the box with the next number in the sequence:

1,884
2,384
2,884

☐

7. $2\left(\sqrt{25} \times \sqrt{25}\right) =$

For Problems 8–9, use the bar graph to the right.

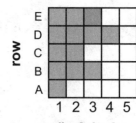

8. According to the graph, how many desks were in row A? _____

9. Which two rows had the same number of desks? _____ and _____

10. What is the remainder after each number is divided?

$9\overline{)76}$ _____ $6\overline{)59}$ _____ $4\overline{)89}$ _____

Sixth-Grade Math Minutes © 2007 Creative Teaching Press

NAME:

MINUTE 62

1. A good runner might be able to run _____ miles in one hour.

    a. 20   **b.** 30   **c.** 10

2. Which of these shapes is a rhombus?

    a.   **b.**   **c.**

3. Use + or × to complete the problem. $\frac{1}{6}\ \square\ \frac{4}{6}=\frac{5}{6}$.

4. $2 \cdot 2 \cdot 2 \cdot 3^2 = 36$ Circle: True   or   False

5. If you add 12 to the quotient of 15 divided by 3, you get _____.

6. The cards to the right were placed facedown on a table and then mixed up. Which letter would be most likely to appear when a card is flipped over? _____

G	R	Q	A
G	C	G	S
G	T	B	G
N	G	L	L
P	N	Q	G

7. Write as an improper fraction: $8\frac{3}{4} =$

8. Write in mixed fraction form: $\frac{9}{5} =$

9. $0.327 \times 100 =$ $0.327 \times 10 =$ $0.327 \times 0.1 =$

10. 10% of 46 = _____   10% of 140 = _____

MINUTE 63

1. Which numbers can both 6 and 12 be evenly divided by? Circle: 2 3 4 6 8 12

2. If ▢ is at (2,3), then ■ is at _____.

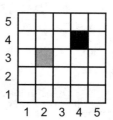

3. If $2^3 = 2 \cdot 2 \cdot 2 = 8$, then $3^3 =$ _____.

4. Below are some perfect square root numbers. What would the next perfect square root be?

$\sqrt{4}$ $\sqrt{9}$ $\sqrt{16}$ $\sqrt{25}$ _____

5. If $3x + 5 = 20$, which of these numbers could x equal?

a. 10   **b.** 15   **c.** 5   **d.** 20

6. The square root of what number is 9? _____

7. What is the perimeter of the shape to the right? _____

For Problems 8–9, use the frequency chart to the right.

8. On which day of the week did Doug mow the most lawns? _____

9. On _____ and _____, Doug mowed the same number of lawns.

Lawns Doug Mowed	
Mowing Day	**Tally**
M	||||
T	||
W	||
TH	|||
F	|
S	|||||
SUN	

For Problem 10, use the rules of negatives to help you simplify each expression.

10.

$(-6)(4) =$

$(-6)(-5) =$

$(7)(-8) =$

Negative × Positive = Negative
Negative × Negative = Positive
Negative + Negative = Negative
Negative ÷ Negative = Positive
Negative ÷ Positive = Negative

Sixth-Grade Math Minutes © 2007 Creative Teaching Press

MINUTE 64

1. Which activity is more likely to occur?
a. getting a hole in one
b. bowling a 300 game

Activity	Odds
hole in one (golf)	33,000 to 1
bowling a 300 game	11,500 to 1

2. What are the coordinates of the ■ ? _____

3. $5\frac{1}{3} + 6\frac{1}{3} =$

4. Fill in the missing factors of 24.

1	2	3		6	8		24

5. Complete the pattern. 1, 3, 7, 15, _____

6. $3 \times (\boxed{} + 4) = 18$

7. Which one of the following is NOT equal to the others?

30%   0.3 $\frac{3}{10}$ 0.03

8. $10^3 =$

9. $(-9) \div (-3) =$ $(-15) \div (3) =$ $(30) \div (-10) =$

For Problem 10, use the rules of negatives to help you simplify each expression.

10. $(-8)(-8) =$
$(9)(-5) =$
$(-7)(9) =$

Negative x Positive = Negative
Negative x Negative = Positive
Negative + Negative = Negative
Negative ÷ Negative = Positive
Negative ÷ Positive = Negative

MINUTE 65

1. Match each word with its definition:

   Prime   **a.** numbers that evenly divide another number

   Factors   **b.** whole numbers that are the products of other numbers

   Multiples   **c.** a number that can only be divided by 1 and itself

For Problems 2–3, use the graph to the right.

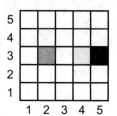

2. What is the distance from one shaded box to the other? _____

3. To get from the gray box to the black box, you would move _____.

   a. north   **b.** south   **c.** east   **d.** west

4. $10 - (6 + 2) =$

5. If $\dfrac{4}{9} = \dfrac{x}{36}$, then $x =$ _____.

6. If $3 + 6 + 2 + 8 + 3 + n = 27$, then $n =$ _____.

For Problems 7–9, circle the greatest amount.

7. 5^3 $\sqrt{25}$ 10^2

8. 3 weeks   20 days   1 month

9. $(-5)(-5)$ $4 \cdot 6$ $\dfrac{100}{5}$

For Problem 10, use the rules of negatives to help you simplify each expression.

10. $(-8) + (-5) =$ $4 - (-5) =$

   | Negative + Negative = Negative |
   | Positive − Negative = Positive |

MINUTE 66

1. Match each kind of fraction with the correct example.

Improper _____   **a.** $\dfrac{5}{4}, \dfrac{4}{5}$

Mixed _____   **b.** $4\dfrac{1}{2}$

Reciprocal _____   **c.** $\dfrac{9}{5}$

2. What is the perimeter of the shape? _____

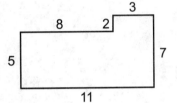

3. $6\dfrac{1}{4} - 5\dfrac{3}{4} =$

4. If $|-6| = 6$, then $|-100| =$ _____.

5. Which numbers can both 8 and 24 be evenly divided by?

Circle: 1   2   3   4   6   8   12

6. Complete the sequence: $\dfrac{1}{8}, \dfrac{1}{4}, \dfrac{3}{8}, \dfrac{1}{2},$ _____.

For Problems 7–10, match each mathematical expression with its correct description.

7. $a + b$ **a.** b is subtracted from a

8. $a - b$ **b.** b is added to a

9. ab **c.** b is multiplied by a

10. $\dfrac{a}{b}$ **d.** a is divided by b

MINUTE 67

1. What is the best estimate of how much of this rectangle is shaded?

  a. $\frac{1}{2}$ **b.** $\frac{1}{3}$ **c.** $\frac{1}{10}$

2. Which of the triangles below is equilateral?

  a.   3 △ 3   **b.**   5 △ 5   **c.**   4 ╱╲ 5
            3   3   6

3. $2\frac{2}{7} = \frac{16}{7}$ Circle: True   or   False

4. If $\frac{3}{5} = \frac{x}{40}$, then $x =$ _____.

5. $48 = 2 \cdot 2 \cdot 2 \cdot 2 \cdot \boxed{}$

6. Write as a mixed fraction: $3.75 =$

7. All of the following equal 10 except:

  $\frac{10^3}{10^2}$ $\sqrt{100}$ 5^2 $|-10|$

8. Put these numbers in order from least to greatest: -5, 7, -2, 8, 0. _____

9. $(-3) + (-8) =$ $(-3) + (8) =$ $(-3) - (8) =$

10. $(-12)(-4) =$ $(-12)(4) =$ $\frac{-12}{4} =$

MINUTE 68

1. What is the best estimate of the part of the rectangle that is shaded?

a. $\dfrac{1}{2}$ b. $\dfrac{1}{8}$ c. $\dfrac{1}{3}$ d. $\dfrac{1}{4}$

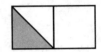

2. Which of the triangles below is isosceles?

a.   b.   c.

3 △ 3   5 △ 5   4 ╲ 5
   3   3   6

3. $0.\overline{3} = $ _____.   a. $\dfrac{1}{2}$ b. $\dfrac{1}{8}$ c. $\dfrac{1}{3}$ d. $\dfrac{1}{4}$

4. $\sqrt{5^2 - 3^2} = $

5. $-12 \div 2 = $ $-12 \div 2 \times (-3) = $

6. $\dfrac{1}{4} \times \boxed{} = 5$

7. Complete the empty boxes.

4
6
8
16
30

$\times \dfrac{1}{2}$

2
3
4

For Problems 8–10, evaluate if $a = 6$, $b = -2$, and $c = -4$.

8. $a + b + c = $

9. $abc = $

10. $a + \dfrac{b}{c} = $

Sixth-Grade Math Minutes © 2007 Creative Teaching Press

NAME: _____

MINUTE 69

1. Which of these is the best estimate of the time on this clock?

    a. noon   **b.** 9:00   **c.** 11:00   **d.** 1:00

2. Which of the triangles below is scalene?

3. Put the following numbers into the correct box below: 3, 14, 2, 4, 21, 6, 8, 28

Multiples of 7

Factors of 24

For Problems 4–6, circle *True* or *False*

4. $(20 \div 2) \cdot 3 = 30$ True   or   False

5. $2(5 + 4) - 6 = 5$ True   or   False

6. $4 + 7 \times 3 = 25$ True   or   False

7. Put the numbers $\{-6, 10, 0, -5, 4\}$ in order from least to greatest. _____

8. Complete the missing numbers in the table.

Sum	Product	Numbers
10	16	2 and 8
8	12	___ and ___

9. $-6 + 8 + 4 - 3 =$ $6 - 8 + 4 - 3 =$

10. $\begin{array}{r} 426 \\ \times\,(-3) \\ \hline \end{array}$ $-3\overline{)513} =$

MINUTE 70

1. $40 \cdot \boxed{\dfrac{}{}} = 10$

2. If point B is halfway between points A and C, what number does it represent? _____

A   B   C

2   16

3. $(3 + 0.3 + 0.7)^2 =$

4. If $4.38 = 4 + \dfrac{a}{10} + \dfrac{8}{b}$, then $a =$ _____ and $b =$ _____.

5. If you spin the spinner to the right, what are the chances it will land on 1 or 3? _____

For Problems 6–9, solve each equation for a.

6. If $a + 8 = 12$, then $a =$ _____.

7. If $a - 2 = -12$, then $a =$ _____.

8. If $-6a = -48$, then $a =$ _____.

9. If $\dfrac{a}{(-3)} = 10$, then $a =$ _____.

10. $\dfrac{1}{4} \times \dfrac{2}{4} =$ $\dfrac{1}{4} + \dfrac{2}{4} =$

MINUTE 71

1. A ton is 2,000 pounds. It might take about _____ sixth graders to weigh a ton.

    a. 25   **b.** 100   **c.** 1,000

2. Match each triangle with its correct definition.

    Equilateral   **a.** a triangle with two equal sides

    Scalene   **b.** a triangle with three equal sides

    Isosceles   **c.** a triangle with no equal sides

3. If Brandon can hop three squares at a time, how many hops will it take him to get to the end of the walkway? _____

4. Put the following numbers into the correct box below: 3, 10, 2, 20, 6, 25

Multiples of 5	Factors of 18

5. If this pattern continues, what letter would be at the top of the next shape in the pattern? _____

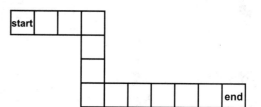

6. Which of these is the same as 7^5?

    a. $7 + 7 + 7 + 7 + 7$ **b.** $5 + 5 + 5 + 5 + 5 + 5 + 5$

    c. $5 \cdot 5 \cdot 5 \cdot 5 \cdot 5 \cdot 5 \cdot 5$ **d.** $7 \cdot 7 \cdot 7 \cdot 7 \cdot 7$

7. Which of these is the same as $0.5888888...$?   **a.** $0.5\overline{8}$ **b.** $\sqrt{0.58}$ **c.** $0.5\overline{8}$ **d.** $|0.58|$

8. Reduce: $\dfrac{5}{15} =$ $\dfrac{10}{24} =$ $\dfrac{6}{30} =$

9. $(-8)(-7) =$ $(-8)(5) =$ $(8)(-4) =$

10. $-5 + (-7) =$ $(-5) - 7 =$ $(-5) - (-7) =$

MINUTE 72

1. Marty got a score of 45 with two throws on this dart board. Which two categories did he hit? _____ _____

30	15
20	10

2. Find the area of one of the triangles. _____

10 / 5 (rectangle with diagonal)

3. Complete the chart.

Fraction	Decimal	Percent
		5%

4. These letters are put on cards and then one card is drawn at random. What is the probability that a Y is drawn? _____

T   T   Y   Y   Y   R   S   S

5. Which of these numbers would solve both of these equations? $2x + 7 = 13$ and $6x - 5 = 13$

a. 3   **b.** 10   **c.** 2

6. All of the following equal 5 except: $|-5|$ $\sqrt{25}$ $\dfrac{5^4}{5^3}$ 5^2

7. If $\dfrac{5}{8} \times a = 1$, then $a = $ _____.

8. $\dfrac{2 \cdot 3 \cdot 3 \cdot 5 \cdot 7}{3 \cdot 5 \cdot 7} = $

(**Hint:** Cross out the common factors in the top and the bottom.)

9. Change to an improper fraction: $4\dfrac{1}{5} = $ $5\dfrac{3}{5} = $ $1\dfrac{9}{10} = $

10. $\left(\dfrac{1}{3}\right)\left(\dfrac{2}{3}\right) = $ $-\left(\dfrac{2}{5}\right)\left(\dfrac{4}{7}\right) = $

MINUTE 73

1. Mike claims he got a score of 55 with two throws on this dart board. Is that possible?   Circle: Yes   or   No

2. Find the area of either right triangle. _____

For Problems 3–4, use the game board to the right.

Red	Red	Blue	Blue
Blue	Red	Blue	Blue
Blue	Red	Red	Blue
Blue	Red	Red	Blue

3. A coin is tossed on the game board. Would it land on a Red or a Blue square more often? _____

4. What is the probability the coin would land on Red?

_____

5. Fill in the missing factors of 28.

1	2		7		28

6. $\dfrac{6 \cdot 5 \cdot 4 \cdot 3 \cdot 2 \cdot 1}{4 \cdot 3 \cdot 2 \cdot 1} =$

7. If $\dfrac{7}{2} \times q = 1$, then $q =$ _____.

8. One of the black squares has the coordinates of (4,5). What coordinates does the other square have? _____

9. If point B is halfway between points A and C, what number does it represent? _____

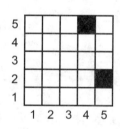

10. Circle the problems below that have a whole number answer.

$400 \div 5$ $\dfrac{300}{10}$ $|-16|$ $\dfrac{1}{4} + \dfrac{1}{4} + \dfrac{1}{4} + \dfrac{1}{4}$

Sixth-Grade Math Minutes © 2007 Creative Teaching Press

MINUTE 74

1. If $\dfrac{5}{8} \div \dfrac{2}{3} = \dfrac{5}{8} \cdot \dfrac{3}{2}$, then $\dfrac{4}{8} \div \dfrac{2}{5} = \dfrac{4}{7} \cdot \boxed{}$

2. To find the volume of a box, multiply all three dimensions. What is the volume of this box? _____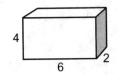

3. What is the common denominator for $\dfrac{1}{4} + \dfrac{1}{5}$? _____

For Problems 4–7, match each clue with its correct answer.

4. the square root of 9   **a.** 20

5. a 9 squared   **b.** 3

6. a factor of 10   **c.** 5

7. a multiple of 10   **d.** 81

For Problems 8–10, evaluate if $a = -5$, $b = -4$, and $c = -3$.

8. $a + b + c =$

9. $abc =$

10. $a - c =$

MINUTE 75

1. How many legs do each of the following have?

4 chairs have _____ legs

5 ducks have _____ legs

2. What is the volume of this box? _____

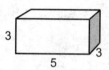

3. 50% + 10% + 0.05 =

4. 20% of 30 is _____.

For Problems 5–7, solve for *x*.

5. If $x - 25 = 96$, then $x =$ _____.

6. If $1.5x = 6$, then $x =$ _____.

7. If $\frac{3}{8}x = 1$, then $x =$ _____.

For Problems 8–9, use the coordinate graph to the right.

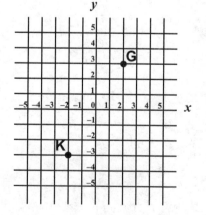

8. What are the coordinates of G? _____

9. What are the coordinates of K? _____

10. $\dfrac{-15}{-3} =$ $(-5)(3) =$ $\dfrac{40}{-5} =$ $(-6)(-3) =$

Sixth-Grade Math Minutes © 2007 Creative Teaching Press

MINUTE 76

1. 1 ton – 300 pounds = _____ pounds.

2. Find the area of the right triangle. _____

3

6

3. Complete the chart.

Fraction	Decimal	Percent
$\dfrac{3}{2}$		

4. Tina wants dark-colored tile for her floor. Which tile has more black squares? _____

Tile A   Tile B

5. $\dfrac{9 \cdot 5 \cdot 7 \cdot 3 \cdot 6 \cdot 0}{4 \cdot 3 \cdot 2 \cdot 1} =$

6. $-3(4 + 5) + 2 =$

x	y
-1	1
-3	-1
-5	-3

7. To get the y number, you add _____ to the x number.

For Problems 8–10, use the coordinate graph to the right.

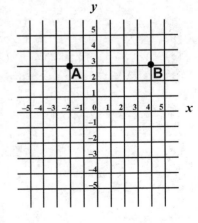

8. What are the coordinates of A? _____

9. What is the distance from A to B? _____

10. To get from B to A, you would travel:
    a. east   **b.** west
    c. north   **d.** south

MINUTE 77

1. Cross out the three-dimensional shape.

2. How many lines of symmetry does this shape have? _____

3. If $a - 13 = -8$, then $a =$ _____ .

4. Complete the sequence: $\dfrac{1}{2}, \dfrac{3}{5}, \dfrac{5}{8}, \dfrac{7}{11},$ _____ , _____ .

5. I am an even number between 30 and 40. If you add my digits together you get 7. What number am I? _____

For Problems 6–8, cross out the number that does not belong in each list.

6.   3   11   13   6

7.   7   8   14   21

8.   131   272   494   126

9.   10% of 60 =   20% of 60 =   30% of 60 =

10.   138.6 ÷ 10 =   13.86 ÷ 100 =   0.1386 ÷ 10 =

Sixth-Grade Math Minutes © 2007 Creative Teaching Press

MINUTE 78

1. A gallon of gas costs $2.93 per gallon. Marcie's car holds 10 gallons. If her tank is empty, how much will it cost to fill it? _____

2. If $x > 3$, which of these numbers could be a possible number for x?

    a. 3   **b.** -22   **c.** 0   **d.** 4

3. $\dfrac{3}{4} \div \dfrac{1}{3} =$

4. All of the following are equal except: 1 $\dfrac{3}{3}$ $\dfrac{-3}{-3}$ $\dfrac{2}{4}$

5. Which of these fractions is not completely reduced? $\dfrac{2}{6}$ $\dfrac{2}{5}$ $\dfrac{3}{7}$

For Problems 6–8 use > , < , or =.

6. $(6)^2$ _____ $(-6)^2$

7. -5 _____ $|-5|$

8. $0.372 \times 1,000$ _____ 37.2×100

9. $\begin{array}{r} 3{,}281 \\ \times\ 7 \\ \hline \end{array}$

10. $6\overline{)11{,}802} =$

MINUTE 79

1. If Hal usually mows 21 yards per week, how many yards does he average per day? _____

2. $\dfrac{3}{7} \div \dfrac{2}{3} =$

3. Which of these is the correct way to write the number 27.36?
  a. Twenty-seven and thirty six tenths
  b. Twenty seven and thirty six hundredths
  c. Twenty-seven and thirty-six hundredths

For Problems 4–7, match each clue with its correct answer.

4. The positive square root of 9.   **a.** 4

5. Nine squared.   **b.** 24

6. A factor of 8.   **c.** 81

7. A multiple of 12.   **d.** 3

For Problems 8–9, use the circle graph and table of information to the right.

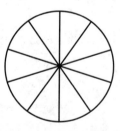

8. The circle has been divided into 10 equal sections. According to the chart, how many sections would need to be shaded for category B? _____

9. How many sections would be shaded for category C?

   _____

Category	Percent
A	10%
B	20%
C	40%
D	30%

10. Complete the chart:

Numbers	Sum	Product	Difference	Quotient
-9, 3				

Sixth-Grade Math Minutes © 2007 Creative Teaching Press

MINUTE 80

1. If two darts were thrown at the board to the right, _____ could be a possible score.   **a.** 15   **b.** 26   **c.** 20

6	2	12
10	8	4

2. The dotted lines represent the lines of symmetry of this shape. What is the perimeter? _____

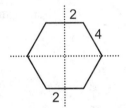

For Problems 3–6, match the correct value of *n*.

3. $n + 6 = -1$ **a.** $n = 5$

4. $-3n = -15$ **b.** $n = -20$

5. $n^2 = 16$ **c.** $n = -7$

6. $\dfrac{n}{-5} = 4$ **d.** $n = 4$

For Problems 7–8, use the graph to the right.

7. This graph shows the value of the stock of a certain company during the first six months of the year. If you bought the stock in January and sold the stock in May, would you have made money or lost money?

_____

8. If you bought the stock in February and sold it in March, would you have made money or lost money?

_____

Stock value per month in dollars

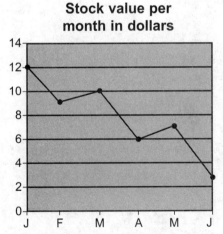

9. $4 \times 0.5 =$ $4 \times 1.5 =$ $4 \times 2.5 =$

10. If $y = 2x + 1$ and $x = 4$, then $y =$ _____.

Sixth-Grade Math Minutes © 2007 Creative Teaching Press

NAME: _____

MINUTE 81

1. Calvin reads an average of eight pages a night. About how many pages will he read in two weeks? _____

2. Match each number with its word:

thirty-eight and six hundredths   **a.** 38.6

thirty-eight and six tenths   **b.** 38.06

three and eight hundred six thousandths   **c.** 3.806

3. Match each statement with its correct answer.

The letter T has _____.   **a.** two obtuse angles and an acute angle

The letter V has _____.   **b.** two right angles

The letter Y has _____.   **c.** an acute angle

For Problems 4–7, circle *True* or *False*.

4. $10 + 32 = 16$ True   or   False

5. $2(5 - 10) + 2 = -8$ True   or   False

6. $\dfrac{4 + 3 - 9}{2} = 1$ True   or   False

7. $-3 + -4 \cdot 2 = -11$ True   or   False

8. Put the following numbers into the correct box below: 3, 8, 15, 10, 2

Factors of 15	Factors of 40

9. In Problem 8, could the number 5 be placed in either box?   Circle: Yes   or   No

10. $\dfrac{1}{10} + \dfrac{6}{10} =$ $\dfrac{1}{10} \times \dfrac{6}{10} =$ $-\left(\dfrac{1}{10}\right) \times -\left(\dfrac{6}{10}\right) =$

Sixth-Grade Math Minutes © 2007 Creative Teaching Press

MINUTE 82

1. Place a decimal point in the following number so that the 3 has a value of $\frac{3}{10}$: 2 4 3 5 9

For Problems 2–4, use the coordinate graph to answer *True* or *False*.

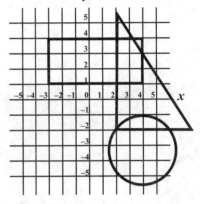

2. The point (3,2) is inside the triangle and rectangle.
Circle: True or False

3. The point (3,-4) is inside the circle.
Circle: True or False

4. The point (-1,3) is outside of all three shapes.
Circle: True or False

5. $\frac{9}{3} + (4 \cdot 2) =$

Letter	Number
A	
B	
C	
D	

For Problems 6–8, use the table and bar graph to the right.

6. Use the graph to help you fill in the table with the number of students who received each grade.

7. According to the graph, there were three times as many _____ as _____.

8. Were there any Fs in Miss Roth's class? _____

Miss Roth's Class Grades

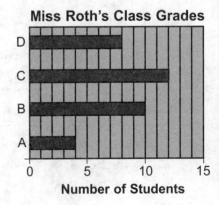

9. Find the area and perimeter of each square.

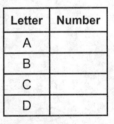 4 cm   Area: _____
Perimeter: _____

8 ft.   Area: _____
Perimeter: _____

10. 2 • 1 =   3 • 2 • 1 =   4 • 3 • 2 • 1 =

Sixth-Grade Math Minutes © 2007 Creative Teaching Press

MINUTE 83

1. How many degrees must the temperature rise to reach the record high? _____

Current Temp.	Record High
83°	91°

2. How many faces does this shape have? _____

3. The top and bottom of the letter Z are _____.
   a. parallel   **b.** perpendicular   **c.** neither

4. List the factors of 12. _____, _____, _____, _____, _____, _____.

5. List the factors of 18. _____, _____, _____, _____, _____, _____.

6. What is the greatest common factor (GCF) that Problems 4 and 5 have in common? _____

7. What should the next shape in the pattern be? _____ ◯ ⬭ ⬭ ◯ ⬭
   a. ⬭   **b.** ◯

8. Ivan has soccer practice at 3:30 and a banquet at 6:00. If soccer practice lasts an hour, how much time will he have to get ready for the banquet? _____

9. Three boxes have the following dimensions. Find their volumes:
   Box 1: 2, 4, 5   Volume = _____ cubic units
   Box 2: 3, 3, 4   Volume = _____ cubic units
   Box 3: 2, 5, 8   Volume = _____ cubic units

10. Circle the prime number in each row.
   5   8   10
   4   12   23
   21   18   29

Sixth-Grade Math Minutes © 2007 Creative Teaching Press

MINUTE 84

1. If the first circle and then every other circle below were shaded, how many would be shaded? _____

◯ ◯ ◯ ◯ ◯ ◯ ◯

For Problems 2–5, use < , >, or = to complete.

2. $3.\overline{8}$ _____ 3.5

3. radius _____ diameter

4. 52 _____ $\sqrt{36}$

5. 1 _____ $|-1|$

6. What is the next shape in the pattern? _____   ◯ ▢ △ ◯ ▢

    a. ◯   **b.** ▢   **c.** △

7. What is the greatest common factor (GCF) of 30 and 40? _____

8. Should the shaded square of the pattern have a dot in it? _____

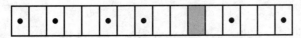

9. Complete each statement with the correct number of angles.

A rectangle has _____ angles.

An octagon has _____ angles.

A hexagon has _____ angles.

10. Complete the chart.

Numbers	Sum	Product	Difference	Quotient
-20, -4				

Sixth-Grade Math Minutes © 2007 Creative Teaching Press

MINUTE 85

1. 76 minutes = _____ hour(s) and _____ minutes.

For Problems 2–3, use the coordinate graph to the right.

2. In which quadrant would the point (-3,5) be found? _____

3. In which quadrant would the point (4,-6) be found? _____

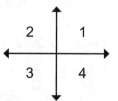

For Problems 4–6, circle the greatest amount.

4. 12%   0.15 $\frac{1}{5}$

5. 4 – (-7)   20 + (-5) $(-3)^2$

6. obtuse angle   acute angle   right angle

7. What is the greatest common factor (GCF) of 18 and 27? _____

8. Which of these four friends has a money amount that could be divided evenly by 3? _____

Naomi	$42
Maria	$50
Barry	$58
Lisa	$65

9. Add the three numbers, and then divide the answer by 3 to get the average.

2, 3, 7   Average = _____
5, 6, 10   Average = _____
2, 4, 9   Average = _____

10. $1.2\overline{)2{,}568}$ =

Sixth-Grade Math Minutes © 2007 Creative Teaching Press

MINUTE 86

1. Jamie planned on splitting her package of candy evenly with her friend Ali. When she opened the package, she found that this was not possible. Which of the following could be the number of pieces of candy in her package?

    a. 12   **b.** 21   **c.** 16   **d.** 20

2. Which of these could NOT be the angles of a triangle?

    a. $100°, 50°, 30°$ **b.** $100°, 50°, 40°$

3. If 3:RED and 4:BLUE, then 6: _____.

    a. GREEN   **b.** BROWN   **c.** ORANGE   **d.** PINK

4. The number 0.2 would best belong between which two of these fractions? $\dfrac{1}{8} \quad \dfrac{1}{4} \quad \dfrac{3}{8} \quad \dfrac{1}{2}$

    a. $\dfrac{1}{8}$ and $\dfrac{1}{4}$ **b.** $\dfrac{1}{4}$ and $\dfrac{3}{8}$ **c.** $\dfrac{3}{8}$ and $\dfrac{1}{2}$

For Problems 5–8, circle *True* or *False*.

5. $32 \cdot 4 = 32$ True   or   False

6. $4\dfrac{2}{5} = \dfrac{22}{5}$ True   or   False

7. $\dfrac{4}{20} = \dfrac{3}{19}$ True   or   False

8. $5\% = 0.5$ True   or   False

For Problems 9–10, evaluate if *a* = 2, *b* = -6, and *c* = 8.

9. $c^a =$

10. $b(a + c) =$

MINUTE 87

1. Jason had $34. He made $15 mowing a lawn. Then he spent $12 golfing. How much money does he have left? _____

2. Study the pattern below. If the pattern continued, what would the sum of the fourth square be? _____

1	2
3	4

First

5	6
7	8

Second

9	10
11	12

Third

3. Linda left for her friend's house at 1:45. Her father told her to be home in 1 hour and 15 minutes. By what time should she be home? _____

4. $\sqrt{\sqrt{16}} =$

5. The weather service predicts a 40% chance of rain for Friday. What is the predicted chance that it won't rain? _____

For Problems 6–9, match each description with its correct expression.

6. Twice a number plus one.   **a.** $n^2 + 1$

7. A number squared plus one.   **b.** $(n + 1)^2$

8. The quantity of n plus 1 squared.   **c.** $2(n + 1)$

9. Twice the quantity of n and 1.   **d.** $2n + 1$

10. $(0.5)(6) =$ $(0.5)(-6) =$ $(0.6)(6) =$ $(-0.6)(-6) =$

MINUTE 88

1. Lynn caught six fish. All of them were between two and three pounds. What was the total weight of all six fish?

    a. between 8 and 9 pounds   **b.** between 12 and 18 pounds

    c. between 20 and 30 pounds

2. Which of these shapes has the greatest perimeter?

    a.   **b.**   **c.**   **d.**

3. What is the total shaded area of all three boxes below as a mixed number? _____

4. If $3n = -60$, then $n =$ _____.

5. If $\dfrac{16}{n} = 8$, then $n =$ _____.

For Problems 6–7, use the coordinate graph to the right.

6. In which quadrant is the point (-3,-4)? _____

7. In which quadrant is the point (-2,2)? _____

For Problems 8–9, use the chart to the right.

8. Jennifer wants to open a bank account with $700. What interest rate will she get for her money? _____

9. What is the minimum amount of money that Tim will need to start a new account? ____

10. What is 1% of $400? _____

General Bank Savings and Loan	
Interest Rate	**Amount**
0%	$20–$199
1%	$200–$499
1.5%	$500–$4,999
2%	$5,000–$9,999
3%	More than $10,000

Sixth-Grade Math Minutes © 2007 Creative Teaching Press

MINUTE 89

For Problems 1–2, use the chart to the right.

Good Numbers
1,331
252
13,531
22

1. Based on the chart, would 2,552 be a good number or a bad number? _____

2. Would 331 be a good number or a bad number? _____

For Problems 3–4, use the calendar to the right.

MAY

S	M	T	W	T	F	S
				1	2	3
4	5	6	7	8	9	10
11	12	13	14	15	16	17
18	19	20	21	22	23	24
25	26	27	28	29	30	31

3. Which day would be two weeks and one day after the shaded one? _____

4. Tammy's birthday is on June 2. What day of the week will this be? _____

5. $\frac{1}{2}(3 \bullet 2) =$

6. Below are five ways the letters HAT can be arranged. What is the sixth way?

HAT   HTA   ATH   AHT   TAH   _____

7. If $a = 11$, then $a^2 =$

8. If $a = -11$, then $a^2 =$

9. $(negative)^2 = positive$ Circle: True   or   False

10. Which of the shaded squares is incorrect on this subtraction problem?   Circle: A   B   C

1	3	5	8
−	4	3	9
	A 9	B 0	C 9

MINUTE 90

1. Randy is talking to his friend in Germany, who says the temperature there is 0° Celsius. This would be closest to what temperature in the United States?
   a. 100° F   **b.** -20° F   **c.** 32° F

2. Which two letters represent the hypotenuse of a triangle in this figure?
   a. AD   **b.** AB   **c.** BC   **d.** BD

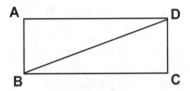

For Problems 3–4, use the grid to the right.

3. Jamie is supposed to shade 25% of the squares. How many more will she need to shade? _____

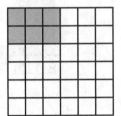

4. What fraction of the grid is currently shaded?

   a. $\dfrac{1}{3}$ **b.** $\dfrac{1}{4}$ **c.** $\dfrac{1}{5}$ **d.** $\dfrac{1}{6}$

5. Which of these letters would look the same if was flipped upside down? _____
      R   A   W   X

6. If the number 35,673 was written backwards, would it be bigger or smaller? _____

7. What number is missing in this sequence? _____   15   12   6   3   0

8. If $7 < a < 11$, then a could equal _____.
   a. 8   **b.** 6   **c.** 12   **d.** 15

9. Reduce: $\dfrac{5}{10} =$ $\dfrac{6}{18} =$

10. Circle the numbers below that are divisible by 2.
   438   537   246   711   25

MINUTE 91

1. Farmer Ed had 11 sheep. All but four of them ran away. How many are left? _____

2. This star has _____.
   a. all acute angles
   b. some acute and some obtuse angles
   c. all obtuse angles

3. What is the total area of all the shaded boxes below as a fraction? _____

4. $\left(3^2\right)^2 =$

5. Fill in the missing factors of 32.

1		4			32

6. How many numbers in the table to the right are prime? _____

3	4	17
7	6	12
8	11	9

7. If $-8 < a < 6$, then a could equal _____.   -5   0   8   -10   1

8. Circle the greatest amount. $\dfrac{1}{9}$ $\dfrac{1}{10}$ 10%   0.06

9. Circle all of the following numbers that are evenly divisible by 5.
   20   35   40   12   10

10. Find the perimeter and area of the right triangle.
   (**Hint:** The longest side is 10 cm.)

   Perimeter = _____   Area = _____

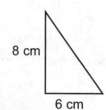

8 cm

6 cm

MINUTE 92

1. $30 \cdot \boxed{} = 6$

2. The volume of the box is 40.
What is the missing dimension? _____

10

2

x

3. Which of these numbers is evenly divisible by both 8 and 6?
    a. 16   **b.** 48   **c.** 32   **d.** 12

For Problems 4–7, solve each equation for _n_.

4. If $n + n + 2 = 10$, then $n = $ _____.

5. If $-6n = -48$, then $n = $ _____.

6. If $\dfrac{n}{12} = \dfrac{15}{36}$, then $n = $ _____.

7. If $\sqrt{n} = 9$, then $n = $ _____.

8. What number is missing in this sequence? 2, -4, _____, -16, 32, …

For Problems 9–10, use the coordinate graph to the right.

9. To get from point A to point B, you must go
_____ and _____.
(up or down) and (right or left)

10. Sandra lives halfway between A and B. Which of these
coordinates describes the location of her house?
    a. (1,2)   **b.** (3,-3)   **c.** (-5,0)

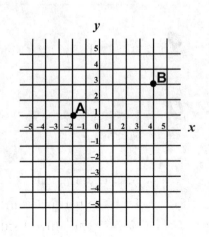

MINUTE 93

1. Vanessa's hens laid 80 eggs today. How many cartons holding a dozen each can she fill completely? _____

2. If the digits in the number 23 are reversed, what is the difference between the original number and the new number? _____

3. If $x = 7$, then $-x =$ _____.

4. Summer school classes begin at 8:30 and last for two and a half hours. At what time do the classes end? _____

5. If $x^3 < 5$, then x could NOT equal: 5   -6   0   -10

6. Fill in the empty boxes.

12	28	63	
21	82		27

7. Would the number $\frac{1}{5}$ be closer to 10%, 25%, or 50%? _____

8. Use the pattern rule to complete the sequence.

| Multiply by 3, then add 1 |

1, 4, 13, 40, _____

9. $6 \cdot 2 + (-3)(4) =$ $2^2 - \sqrt{25} =$

10. Complete the crossword using the clues.

Across
1. $12 \cdot 4 =$
3. One and a half dozen is _____.

Down
2. $9^2 =$
4. 8 dimes = _____ cents.

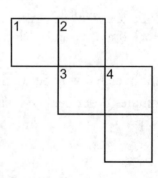

MINUTE 94

1. If today is Sunday, three days ago was _____.

2. Ken paid $30 for a jacket that was 50% off. What was the original price? _____

3. The answer to $\sqrt{28}$ is a _____. Circle: decimal   or   whole number

4. If $x = -5$, then $-x =$ _____.

5. Fill in the empty boxes.

$\frac{3}{4}$	$\frac{5}{11}$	
$\frac{4}{3}$		$\frac{15}{4}$

6. Which of the following are common factors for the numbers 20 and 30?

2   4   5   6   10   15   20

For Problems 7–9, use the grid to the right.

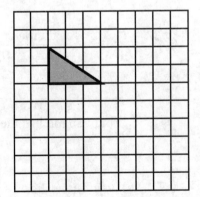

7. If the right triangle's dimensions are enlarged three times, the new base and height would be _____ units and _____ units.

8. What would the area of the enlarged triangle be? _____

9. The hypotenuse (the longest side) of the enlarged triangle would be:

a. greater than 3   **b.** less than 3   **c.** equal to 3

10.   14 + (-10) =   14 − (-10) =   14 • (-10) =

MINUTE 95

1. If a and b are odd whole numbers, which of the following would also be an odd whole number?

   a. ab **b.** $a + b$ **c.** $\dfrac{a}{b}$

2. In the fraction $\dfrac{1}{8}$, 1 is called the _____ and 8 is called the _____.

3. Is $\sqrt{37}$ closer to 6 or 7? _____

4. $-(6 + 5) =$

5. $-(-8 + 4) =$

6. $-(-2) =$

7. If $2n > 12$, then n could equal _____.

   a. 4   **b.** 5   **c.** 6   **d.** 7

8. Which shape has the greater area? _____

   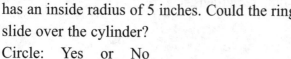

9. The cylinder has a diameter of 9 inches. The ring has an inside radius of 5 inches. Could the ring slide over the cylinder?

   Circle: Yes or No

   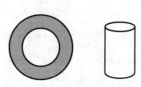

10. Circle the numbers below that are evenly divisible by 4.

    48   505   408   600   102

Sixth-Grade Math Minutes © 2007 Creative Teaching Press

MINUTE 96

1. Three months ago, Janelle weighed 90 pounds. If she has gained an average of four pounds per month, what does she weigh now? _____

2. When you divide fractions, you actually flip the _____ fraction and then multiply.   Circle: first   or   second

3. How many one-inch cubes can be placed in this four-inch cube? _____

4

For Problems 4–5, use the chart to the right.

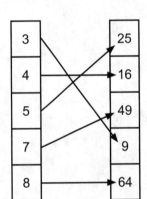

3		25
4		16
5		49
7		9
8		64

4. Why do these numbers have arrows drawn between them? _____

_____

5. Which number in the first column could have gone in the second column? _____

6. If $2a - 4 = a + 1$, then $a = $ _____.
   a. 6   **b.** 5   **c.** 4

7. $\dfrac{1}{2} \times \dfrac{1}{4} \times \dfrac{4}{3} = $

For Problems 8–10, use the diagram to the right.

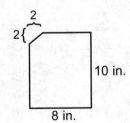

8. This piece of paper was 8 inches by 10 inches before the corner was torn off. What was the area of the paper before the corner got torn off? _____

9. What is the area of the corner (triangle) that got torn off? _____
(**Hint:** The corner was a right triangle that had two-inch legs.)

10. What is the actual area of the paper without the corner? _____

MINUTE 97

1. Abraham Lincoln was born in 1809 and died in 1865. For how many years did he live? _____

2. When you divide fractions, you should _____ the first fraction by the reciprocal of the second fraction.

    a. add   **b.** subtract   **c.** multiply   **d.** divide

3. If $\frac{1}{2}x = 6$, then $x =$ _____.

4. $2 + 4 \cdot \boxed{} = 22$

5. If the pattern continues, what number should be at the top of the steps? _____

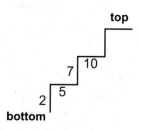

6. If $x = -100$, then $-x =$ _____.

7. If $4a > 11$, then $a =$ _____.

    a. -2   **b.** 2   **c.** -3   **d.** 3

8. In order for the scale to balance, x would have to equal _____.

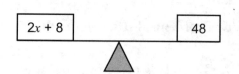

For Problems 9–10, rewrite each problem using exponents.

9. $3 \cdot 3 \cdot 3 \cdot 3 \cdot 2 \cdot 2 =$

10. $5 \cdot 5 \cdot 5 \cdot x \cdot x \cdot y \cdot y \cdot y =$

Sixth-Grade Math Minutes © 2007 Creative Teaching Press

MINUTE 98

For Problems 1–3, use the multiplication problem to the right.
Circle *True* or *False*.

$$\frac{3}{6} \times \frac{6}{12} =$$

1. To simplify this problem, you can cancel the 6s (diagonally).   True   or   False

2. To simplify this problem, you could also reduce $\frac{3}{6}$ to $\frac{1}{2}$.   True   or   False

3. The final answer to this problem would be $\frac{1}{3}$.   True   or   False

4. This shape is divided into _____.
   a. fourths   **b.** thirds
   c. three parts   **d.** triangles

5. Shade the odd multiples of 7.

7	12	14	18	21	28	35

6. Use the numbers 1, 2, 3, and 4 to fill in these boxes and make a correct equation.

☐ + ☐ = ☐ + ☐

7. Circle the fractions that are more than $\frac{1}{2}$. $\frac{3}{10}$ $\frac{3}{5}$ $\frac{2}{3}$ $\frac{2}{4}$ $\frac{5}{9}$

For Problems 8–10, use the diagram and chart.

8. There is one road between towns A and C, as shown on the diagram. What is the distance between towns A and C by road? _____

9. Sally lives in town A. On Saturday, she made a round-trip bike ride to town B. How far did she ride? _____

10. If the bike ride took Sally two hours, solve this proportion to find her average speed in miles per hour.

If $\frac{16}{2} = \frac{x}{1}$, then $x =$ _____.

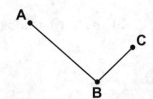

A	B	8 miles
B	C	4 miles

 MINUTE 99

For Problems 1–3, use the division problem to the right.
Circle *True* or *False*.

$$\frac{1}{8} \div \frac{3}{4} =$$

1. To solve this problem, you should rewrite it as $\frac{1}{8} \times \frac{4}{3}$.   True   or   False

2. When dividing fractions, flip the first and multiply by the second.   True   or   False

3. The final answer to this problem would be $\frac{1}{6}$.   True   or   False

For Problems 4–6, use the coordinate graph to the right.

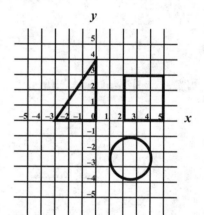

4. What is the area of the triangle? _____

5. What is the area of the square? _____

6. Does the circle or the square have the greatest area?
_____

7. If $\frac{1}{3}a = 1$, then $a = $ _____.

For Problems 8–10, evaluate if *a* = -3, *b* = -12, and *c* = 6.

8. $\frac{c^2}{b} =$

9. $-2(a + b) =$

10. $\frac{c}{b} \cdot \frac{b}{a} =$

MINUTE 100

1. 5,000 tennis balls might fill up a _____.
   a. car   **b.** house   **c.** school

2. What is the radius of this circle if the diameter is 12.6 cm? _____

12.6 cm

3. If $x < 3.4$, which of the following could be a value of x?
   a. 4.6   **b.** 2.8   **c.** 5.1

4. Complete the table.

Fraction	Decimal	Percent
$\frac{3}{5}$		

5. $\dfrac{\sqrt{3^2 + 4^2}}{5} =$

6. If $b - 4.25 = 8.25$, then $b =$ _____.

For Problems 7–9, use the graph to the right.

7. Jason received the same scores on Test ____ and Test ____.

8. Which of these numbers would be closest to Jason's average score?
   a. 93   **b.** 72   **c.** 81

9. If there were 50 questions on Test 1, how many did Jason answer correctly? _____

Jason's Test Scores

10. $\dfrac{-250}{-5} =$ $(-3) + (-4) - (5) =$

Sixth-Grade Math Minutes © 2007 Creative Teaching Press

MINUTE ANSWER KEY

MINUTE 1
1. 49
2. ———
3. 0.034, 0.340, 0.403
4. 3/10
5. 7
6. 17
7. 12 sq. units
8. 5
9. 36, 63, 81
10. 4, 6, 9

MINUTE 2
1. d
2. b
3. 2/5, 3/4
4. 7/10
5. 16
6. 20
7. 14 units
8. A = 5, B = 20, C = 30
9. 48, 32, 56
10. 4, 6, 3

MINUTE 3
1. 5:56
2. 6
3. 2/8, 1/4
4. <
5. 12
6. 16
7. 50 ft.
8. 20 people
9. 36, 60, 72
10. 10, 11, 9

MINUTE 4
1. 41.5
2. c
3. >
4. <
5. 22
6. 456
7. Yes
8. A
9. 6, 12, 18
10. 75, 139, 83

MINUTE 5
1. a
2. D
3. 10
4. 2.3
5. 7 boxes
6. 18
7. 9 sq. units
8. 3
9. 21, 19
10. 70, 161

MINUTE 6
1. c
2. C
3. a, b
4. 0.23
5. 9 miles
6. D24, E28
7. 63 ft.^{2}
8. Thursday
9. Tuesday and Friday
10. 54, 45, 35

MINUTE 7
1. d
2. A
3. a, c
4. 0.043
5. True
6.
7. 18 units
8. Desiree
9. Rick
10. 212, 43, 167

MINUTE 8
1. $5.42
2. A
3. B
4. 4/9, 4/16 or 1/4
5. 4/11
6. 16
7. December, January
8. December
9. 2.9, 4.3, 12.4
10. 88, 170, 276

MINUTE 9
1. 20, 310, 110
2. c
3. a
4. a
5. 4
6. 4/5
7. A
8. Red
9. 696 pounds
10. 0.72, 0.98, 2.08

MINUTE 10
1. c
2. a
3. c
4. 3.5, 5.1
5. 4
6. 4/9
7. B
8. 50 eggs
9. 75 eggs
10. 5.7, 10.1, 17.5

MINUTE 11
1. 4,321
2. d
3. b
4. 2/8, 3/8, 7/8, 8/8
5. 6
6. 27
7. 12 cubes
8. A = 20, B = 25, C = 45
9. 63, 64, 42
10. 15, 17, 19

MINUTE 12
1. b
2. C
3. a
4. 2
5. 14 yards
6. 2 out of 40, or 5%
7. 30 units
8. 30 glasses
9. 5.8, 8.3
10. 56, 63

MINUTE 13
1. 100, 2,300, 0
2. C
3. B
4. 2/6
5. 12/25
6. 27 boys
7. True
8. 1/7
9. 1/6, 1/12, 1/30
10. 30, 63, 72

MINUTE 14
1. 4, 8
2. a
3. 5/15
4. 10
5. 9
6. 1/4
7. a and c
8. b
9. 2.5, 3.25, 20.5
10. 5, 6, 5

MINUTE 15
1. $1.00
2. ┼
3. c
4. >
5. =
6. 2
7. 20 ft.
8. 5
9. 125, 150, 250
10. 31, 102

MINUTE 16
1. 541
2. B
3. a
4. d
5. 4
6. 21
7. 45 ft.^{2}
8. Class 2
9. 5 more girls
10. 1.2, 13.05, 3.5

MINUTE 17
1. $2.67
2. c
3. >
4. >
5. 9
6. B
7. 3/9 or 1/3
8. Jared
9. Jackie
10. 250, 125, 250

MINUTE 18
1. a
2. d
3. 5.60, 5.06, 0.56, 0.056
4. 2/6, 3/9
5. No
6. A
7. 2/5
8. 52
9. 26, 25, 29
10. 5, 9, 11

MINUTE 19
1. b
2.
3. 4/15
4. 1/2
5. less
6. 4
7. 40
8. 7,200
9. 63, 150, 36
10. 7, 7, 3

MINUTE 20
1. b
2. a
3. 6/35
4. 5 people
5. 4 people
6. 3 people
7. 7
8. I
9. 7.5, 11.2, 22.9
10. 30, 12, 70

MINUTE ANSWER KEY

MINUTE 21
1. a
2. b
3. 20
4. 1/24
5. 2
6. Add 5, subtract 1
7. 11 sq. units
8. 5
9. 614
10. 3,301

MINUTE 22
1. 4:48
2. E
3. 3
4. 4/35
5. 64
6. Adding the first two
7. 30 cm
8. 13
9. 4, 6, 3
10. 102, 224

MINUTE 23
1. 1,000, 2,000, 3,000
2. H
3. 3
4. 2
5. 17
6. 40, 60
7. 5
8. 22
9. 54, 72, 81
10. 36, 48, 35

MINUTE 24
1. $2.70
2. a
3. b
4. 5/7
5. 21
6. 10
7. 9 feet
8. 10
9. 860, 930
10. 2,500, 3,600

MINUTE 25
1. $8
2. c
3. c
4. 11/7 or 1 4/7
5. ×
6. 6 sides
7. 48 inches
8. 18
9. 2, 2
10. 6, 9

MINUTE 26
1. 14
2. b
3. b
4. 13/2
5. 50
6. A
7. 4.5 sq. units
8. 56
9. 60, 500
10. 15,087

MINUTE 27
1. 3 quarters, 1 dime, 2 pennies
2. c
3. 11/12
4. 16
5. 2
6. 4, 2
7. 8 sq. units
8. 24 eggs
9. 8,121
10. 6,239

MINUTE 28
1. 36 cookies
2. c
3. 5 2/3
4. 3
5. +
6. a
7. 42
8. 3 students
9. Bs
10. 1.2, 2.8, 4

MINUTE 29
1. Tuesday
2. f
3. 25/3
4. >
5. ×
6. W
7. 7 sq. units
8.
9. 1.9, 1.7
10. 40, 20, 0

MINUTE 30
1. 6th of June
2. 13
3. 4/27
4. 35
5. 65
6. 48
7. 22 units
8. 8
9. =
10. 1/20, 2/21, 3/40

MINUTE 31
1. Wednesday
2. a
3. 75
4. 3/4
5. circle
6. 5
7. 3, 8
8. 2/5
9. 365, 270
10. 309, 247

MINUTE 32
1. Yes
2. b
3. 3/5
4. 90%
5. 10%
6. circle
7. 6
8. 20
9. 2
10. 3/5, 2/25

MINUTE 33
1. 10 weeks
2. 9 quarters
3. 50 inches
4. a
5. c
6. Sally
7. triangle
8. 12
9. 8
10. 40

MINUTE 34
1. a
2. rhombus = b, square = c, quadrilateral = a (also b, c)
3. 39%
4. 11 more boxes
5. 2
6. 5,694,600
7. c
8. 15
9. 23.6, 34, 460
10. 20, 25

MINUTE 35
1. c
2. d
3. 75%
4. 1/2
5. 95%
6. 8
7. 6
8. b
9. 28, 42, 60
10. 1/9, 2/3, 0

MINUTE 36
1. b
2. 2
3. 7/8
4. 1/2
5. 13
6. 4
7. 6
8. 7/35
9. 8
10. 13, 40 and 16, 63

MINUTE 37
1. a
2. Line = b, Segment = a, Ray = c
3. 1/4
4. d
5. c
6. 15
7. a
8. b
9. April
10. 5.62, 42.6, 0.58

MINUTE 38
1. c
2. d
3. 7/12
4. 15/48
5. 10%
6. 40 squares
7. 6
8. Justine
9. 49, 64, 36
10. 10, 10, 10

MINUTE 39
1. d
2. 5, 8, 10
3. 0.55
4. a
5. a
6. The 1 should be an 8.
7. She found the area.
8. 9
9. 24
10. 15

MINUTE 40
1. c
2. 14/20 = 7/10
3. 0.61
4. 47%
5. 29
6. 5
7. 7
8. 33
9. 0.06
10. 300

MINUTE ANSWER KEY

MINUTE 41
1. 1,249
2. B
3. A
4. 0.75, 75% and 1/10, 10%
5. 10
6. 21
7. 4
8. a
9. c
10. d

MINUTE 42
1. b
2. 3/8
3. 50%
4. 1/4, 0.25 and 3/10, 30%
5. 81
6. 32
7. d
8. 18
9. 1/2, 3/64
10. 200, 150, 80

MINUTE 43
1. a
2. Acute, Right, Obtuse
3. d
4. 15
5. carpet
6. inches
7. 2
8. 15, 90
9. 10, 6
10. 38

MINUTE 44
1. c
2. 6 faces
3. $1.\overline{7}$
4. 50%
5. 3
6. b
7. a
8. c
9. 1/3, 11/9 or 1 2/9
10. 19/2, 41/4

MINUTE 45
1. 10 cans
2. 7 faces
3. $0.8\overline{2}$
4. 0.25
5. 10
6. D D D D
7. Grade 4
8. Grade 5
9. 0.111, 0.151
10. 10, 100, 1,000

MINUTE 46
1. 6,543.21
2. 7 units
3. $0.3\overline{9}$
4. 30
5. 55%
6. 25
7. 7
8. A = 12, B = 27, C = 18
9. 1/5, 1/100
10. 0.3, 0.28

MINUTE 47
1. 1,234.56
2. 3
3. 35
4. seventeen hundredths
5. 0.7 or $\frac{7}{10}$
6. 45%
7. 30%
8. c
9. 9.8, 98, 980
10. 5%, 15%, 85%

MINUTE 48
1. d
2. Yes
3. 0.9
4. $0.\overline{7}$
5. 1
6. 0.6
7. 90
8. 106
9. 42, 90, 45
10. 200/10, 0.16 × 100

MINUTE 49
1. 50 mph
2. Yes
3. $0.\overline{7}$
4. 2/5, 0.4 and 0.25, 25%
5. 6
6. 6
7. 3
8. EG
9. 2/10, 4/20
10. 14, 365, 38

MINUTE 50
1. 40 cartons
2. 8 faces
3. <
4. <
5. >
6. 7, 2, 2
7. No
8. $13
9. 15, 6, 6
10. 4, 2, 10

MINUTE 51
1. Joanne: 10, Jackie: 13
2. 180 degrees
3. $0.3\overline{8}$
4. 4
5. Yes
6. No
7. A
8. B
9. 4, 5, 8
10. 16/3, 20/3, 13/4

MINUTE 52
1. b
2. 40°
3. 0.95, 95%, or $\frac{19}{20}$
4. 5
5. CBA
6. 124
7. 5
8. 25
9. c
10. 18 mm², 18 mm

MINUTE 53
1. 165 miles
2. 180
3. 0.3
4. 6
5. 24
6. 7
7. clear
8. 10 dimes
9. 39, 47, 68
10. 36, 30

MINUTE 54
1. $44
2. 10 sq. units
3. 8
4. 54
5. outside
6. 5
7. (3 + 9) × 4 = 48
8. 3.25
9.
10. 15, 15, 15

MINUTE 55
1. 5,650 and 6,000
2. 24 sq. units
3. 3.06, 3.068
4. 8 × 8 = 64
5. 8
6. 1/8
7. 5, 17, 29
8. =
9. >
10. >

MINUTE 56
1. b
2. May 13
3. B = 2.4, C = 2.8, A = 2.1
4. 5 and 23
5. 2/5
6. b
7. c
8. d
9. a
10. b

MINUTE 57
1. 12 pounds
2. 9
3. 6
4. 32
5. 3/5
6. d
7. c
8. 3, 7 or 1, 9
9. 3, 16
10. 4 or -4

MINUTE 58
1. 10 combinations
2. 5 ft.
3. 35 ft.²
4. 100
5. 28
6. ④ ⑧ ⑥
7. 18 people
8. 12, 2
9. 7,430
10. 4/5, 3/25

MINUTE 59
1. 23
2. 5 paychecks
3. 20 units
4. 6
5. 10/17
6. c
7. 3 squares
8. 14
9. 36
10. 1

MINUTE 60
1. a
2. 1/2
3. a
4. 13
5. 27 girls
6. 27
7. d
8. b
9. a
10. c

MINUTE ANSWER KEY

MINUTE 61
1. 130, 3,000, 488
2. 6 cubes
3. 4
4. 5
5. 1/10 or 10%
6. 3,384
7. 50
8. 1
9. B and E
10. 4, 5, 1

MINUTE 62
1. c
2. b
3. +
4. False
5. 17
6. G
7. 35/4
8. 1 4/5
9. 32.7, 3.27, 0.0327
10. 4.6, 14

MINUTE 63
1. 2, 3, 6
2. (4,4)
3. $3 \cdot 3 \cdot 3 = 27$
4. $\sqrt{36}$
5. c
6. 81
7. 12 units
8. Saturday
9. Tuesday and Wednesday
10. -24, 30, -56

MINUTE 64
1. b
2. (4,3)
3. 11 2/3
4. 4, 12
5. 31
6. 2
7. 0.03
8. 1,000
9. 3, -5, -3
10. 64, -45, -63

MINUTE 65
1. Prime = c, Factors = a,
   Multiples = b
2. 3 units
3. c
4. 2
5. 16
6. 5
7. 5^3
8. 1 month
9. (-5)(-5)
10. -13, 9

MINUTE 66
1. Improper = c, Mixed = b,
   Reciprocal = a
2. 36 units
3. 1/2
4. 100
5. 1, 2, 4, 8
6. 5/8
7. b
8. a
9. c
10. d

MINUTE 67
1. b
2. a
3. True
4. 24
5. 3
6. 3 3/4
7. 5^2
8. -5, -2, 0, 7, 8
9. -11, 5, -11
10. 48, -48, -3

MINUTE 68
1. d
2. b
3. c
4. 4
5. -6, 18
6. 20
7. 8, 15
8. 0
9. 48
10. 6 1/2

MINUTE 69
1. a
2. c
3. Multiples of 7: 14, 21, 28
   Factors of 24: 3, 2, 4, 6, 8
4. True
5. False
6. True
7. -6, -5, 0, 4, 10
8. 2 and 6
9. 3, -1
10. -1,278, -171

MINUTE 70
1. 1/4
2. 9
3. 16
4. $a = 3, b = 100$
5. 2/3
6. 4
7. -10
8. 8
9. -30
10. 1/8, 3/4

MINUTE 71
1. a
2. Equilateral = b, Scalene = c,
   Isosceles = a
3. 4 hops
4. Multiples of 5: 10, 20, 25
   Factors of 18: 3, 2, 6
5. A
6. d
7. c
8. 1/3, 5/12, 1/5
9. 56, -40, -32
10. -12, -12, 2

MINUTE 72
1. 30 and 15
2. 25 sq. units
3. 1/20, 0.05
4. 3/8
5. a
6. 5^2
7. 8/5
8. 6
9. 21/5, 28/5, 19/10
10. 2/9, -(8/35)

MINUTE 73
1. No
2. 18 sq. units
3. Blue
4. 7/16
5. 4, 14
6. 30
7. 2/7
8. (5,2)
9. 7
10. All of them

MINUTE 74
1. 5/2
2. 48 cubic units
3. 20
4. b
5. d
6. c
7. a
8. -12
9. -60
10. -2

MINUTE 75
1. 16, 10
2. 45 cubic units
3. 0.65 or 65%
4. 6
5. 121
6. 4
7. 8/3
8. (2,3)
9. (-2,-3)
10. 5, -15, -8, 18

MINUTE 76
1. 1,700
2. 9 sq. units
3. 1.5, 150%
4. Tile A
5. 0
6. -25
7. 2
8. (-2,3)
9. 6 units
10. b

MINUTE 77
1.
2. 1
3. 5
4. 9/14, 11/17
5. 34
6. 6
7. 8
8. 126
9. 6, 12, 18
10. 13.86, 0.1386, 0.01386

MINUTE 78
1. $29.30
2. d
3. 9/4 or 2 1/4
4. 2/4
5. 2/6
6. =
7. <
8. <
9. 22,967
10. 1,967

MINUTE 79
1. 3 yards
2. 9/14
3. c
4. d
5. c
6. a
7. b
8. 2 sections
9. 4 sections
10. -6, -27, -12, -3

MINUTE 80
1. c
2. 24 units
3. c
4. a
5. d
6. b
7. lost money
8. made money
9. 2, 6, 10
10. 9

MINUTE ANSWER KEY

MINUTE 81
1. 112 pages
2. b, a, c
3. b, c, a
4. False
5. True
6. False
7. True
8. Factors of 15: 3, 15
   Factors of 40: 8, 10, 2
9. Yes
10. 7/10, 3/50, 3/50

MINUTE 82
1. 24.359
2. True
3. True
4. False
5. 11
6. A = 4, B = 10, C = 12, D = 8
7. Cs, As
8. No
9. Area 1: 16 cm^2
   Perimeter 1: 16 cm
   Area 2: 64 ft.^{2}
   Perimeter 2: 32 ft.
10. 2, 6, 24

MINUTE 83
1. 8 degrees
2. 10 faces
3. a
4. 1, 2, 3, 4, 6, 12
5. 1, 2, 3, 6, 9, 18
6. 6
7. a
8. 1 hour 30 minutes
9. 40, 36, 80
10. 5, 23, 29

MINUTE 84
1. 4 circles
2. >
3. <
4. >
5. =
6. c
7. 10
8. Yes
9. 4, 8, 6
10. -24, 80, -16, 5

MINUTE 85
1. 1 hour 16 minutes
2. Quadrant 2
3. Quadrant 4
4. 1/5
5. 20 + (-5)
6. obtuse angle
7. 9
8. Naomi

9. 4, 7, 5
10. 2,140

MINUTE 86
1. b
2. b
3. c
4. a
5. False
6. True
7. False
8. False
9. 64
10. -60

MINUTE 87
1. $37
2. 58
3. 3:00
4. 2
5. 60%
6. d
7. a
8. b
9. c
10. 3, -3, 3.6, 3.6

MINUTE 88
1. b
2. d
3. 2 5/9
4. -20
5. 2
6. Quadrant 3
7. Quadrant 2
8. 1.5%
9. $20
10. $4

MINUTE 89
1. good (first and last digits are 2)
2. bad (first digit is not 1)
3. Wednesday
4. Monday
5. 3
6. THA
7. 121
8. 121
9. True
10. B

MINUTE 90
1. c
2. d
3. 3 squares
4. d
5. X
6. bigger
7. 9
8. a
9. 1/2, 1/3
10. 438, 246

MINUTE 91
1. 4 sheep
2. b
3. 3 1/4
4. 81
5. 2, 8, 16
6. 4
7. -5, 0, 1
8. 1/9
9. 20, 35, 40, 10
10. Perimeter = 24 cm, Area = 24 cm^2

MINUTE 92
1. 1/5
2. 2
3. b
4. 4
5. 8
6. 5
7. 81
8. 8
9. up and right
10. a

MINUTE 93
1. 6 cartons
2. 9
3. -7
4. 11:00
5. 5
6. 36, 72
7. 25%
8. 121
9. 0, -1
10. Across: 1. 48; 3. 18
    Down: 2. 81; 4. 80

MINUTE 94
1. Thursday
2. $60
3. decimal
4. 5
5. 11/5, 4/15
6. 2, 5, 10
7. 9, 6
8. 27 sq. units
9. a
10. 4, 24, -140

MINUTE 95
1. a
2. numerator, denominator
3. 6
4. -11
5. 4
6. 2
7. d
8. square
9. Yes
10. 48, 408, 600

MINUTE 96
1. 102 pounds
2. second
3. 64
4. The second-column numbers are the squares of the first-column numbers.
5. 4
6. b
7. 1/6
8. 80 sq. inches
9. 2 sq. inches
10. 78 sq. inches

MINUTE 97
1. 56 years
2. c
3. 12
4. 5
5. 15
6. 100
7. d
8. 20
9. $3^4 \cdot 2^2$
10. $5^3 \cdot x^2 \cdot y^3$

MINUTE 98
1. True
2. True
3. False
4. c
5. 7, 21, 35
6. 2 + 3 = 4 + 1 (order may vary)
7. 3/5, 2/3, 5/9
8. 12 miles
9. 16 miles
10. 8 miles per hour

MINUTE 99
1. True
2. False
3. True
4. 6 sq. units
5. 9 sq. units
6. square
7. 3
8. -3
9. 30
10. -2

MINUTE 100
1. a
2. 6.3 cm
3. b
4. 0.6, 60%
5. 1
6. 12.5
7. 1 and 4
8. c
9. 40 questions
10. 50, -12